Algebra through practice

Book 3: Groups, rings and fields

Algebra through practice

A collection of problems in algebra with solutions

Book 3

Groups, rings and fields

T.S.BLYTH ○ E.F.ROBERTSON

University of St Andrews

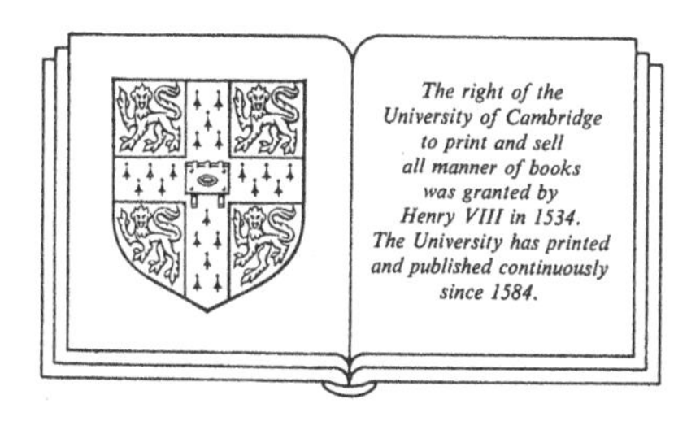

CAMBRIDGE UNIVERSITY PRESS

Cambridge

London New York New Rochelle

Melbourne Sydney

CAMBRIDGE UNIVERSITY PRESS
Cambridge, New York, Melbourne, Madrid, Cape Town, Singapore, São Paulo

Cambridge University Press
The Edinburgh Building, Cambridge CB2 8RU, UK

Published in the United States of America by Cambridge University Press, New York

www.cambridge.org
Information on this title: www.cambridge.org/9780521272889

First published 1984
Re-issued in this digitally printed version 2008

A catalogue record for this publication is available from the British Library

Library of Congress Catalogue Card Number: 83-24013

ISBN 978-0-521-27288-9 paperback

Contents

Preface

The aim of this series of problem-solvers is to provide a selection of worked examples in algebra designed to supplement undergraduate algebra courses. We have attempted, mainly with the average student in mind, to produce a varied selection of exercises while incorporating a few of a more challenging nature. Although complete solutions are included, it is intended that these should be consulted by readers only after they have attempted the questions. In this way, it is hoped that the student will gain confidence in his or her approach to the art of problem-solving which, after all, is what mathematics is all about.

The problems, although arranged in chapters, have not been 'graded' within each chapter so that, if readers cannot do problem n this should not discourage them from attempting problem $n + 1$. A great many of the ideas involved in these problems have been used in examination papers of one sort or another. Some test papers (without solutions) are included at the end of each book; these contain questions based on the topics covered.

TSB, EFR
St Andrews

Background reference material

Courses on abstract algebra can be very different in style and content. Likewise, textbooks recommended for these courses can vary enormously, not only in notation and exposition but also in their level of sophistication. Here is a list of some major texts that are widely used and to which the reader may refer for background material. The subject matter of these texts covers all six of the present books, and in some cases a great deal more. For the convenience of the reader there is given overleaf an indication of which parts of which of these texts is most relevant to the appropriate chapters of this book.

[1] I. T. Adamson, *Introduction to Field Theory*, Cambridge University Press, 1982.

[2] F. Ayres, Jr, *Modern Algebra*, Schaum's Outline Series, McGraw-Hill, 1965.

[3] D. Burton, *A First Course in Rings and Ideals*, Addison-Wesley, 1970.

[4] P. M. Cohn, *Algebra*, Vol. I, Wiley, 1982.

[5] D. T. Finkbeiner II, *Introduction to Matrices and Linear Transformations*, Freeman, 1978.

[6] R. Godement, *Algebra*, Kershaw, 1983.

[7] J. A. Green, *Sets and Groups*, Routledge and Kegan Paul, 1965.

[8] I. N. Herstein, *Topics in Algebra*, Wiley, 1977.

[9] K. Hoffman and R. Kunze, *Linear Algebra*, Prentice Hall, 1971.

[10] S. Lang, *Introduction to Linear Algebra*, Addison-Wesley, 1970.

[11] S. Lipschutz, *Linear Algebra*, Schaum's Outline Series, McGraw-Hill, 1974.

[12] I. D. Macdonald, *The Theory of Groups*, Oxford University Press, 1968.

[13] S. MacLane and G. Birkhoff, *Algebra*, Macmillan, 1968.

[14] N. H. McCoy, *Introduction to Modern Algebra*, Allyn and Bacon, 1975.

[15] J. J. Rotman, *The Theory of Groups: An Introduction*, Allyn and Bacon, 1973.

[16] I. Stewart, *Galois Theory*, Chapman and Hall, 1973.
[17] I. Stewart and D. Tall, *The Foundations of Mathematics*, Oxford University Press, 1977.

References useful to Book 3

1: Groups [2, Chapter 9], [6, Chapter 7], [7, Chapters 4, 5, 6], [8, Chapter 2], [14, Chapters 7, 8].
2: Rings and fields [2, Chapters 10, 11, 12], [6, Chapter 8], [8, Chapter 3], [14, Chapters 2, 3, 9, 10].

In [2, 7, 14] the authors write mappings on the right and, as a consequence multiply permutations accordingly. In contrast, in [8] most mappings are written on the left, but permutations are written and multiplied as mappings on the right. Also in [8] an integral domain is not required to have a 1. In [6] integral domains are not required to be commutative, and all rings have a 1.

1: Groups

Abstract algebra is basically a study of sets with binary operations. A binary operation (or law of composition) on a set E is a mapping $f: E \times E \to E$ described variously by $(x, y) \to x * y$, $(x, y) \to x + y$, $(x, y) \to xy$, etc. When E is finite it is sometimes convenient to represent a binary operation on E by means of a Cayley table, the interpretation of which is that $x_i * x_j$ appears at the intersection of the ith row and the jth column (Fig. 1.1). A binary operation

Fig.1.1

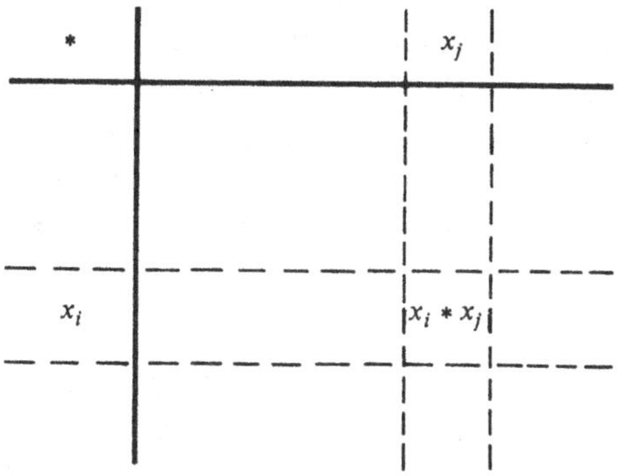

$*$ on E is associative if $(\forall x, y, z \in E) x * (y * z) = (x * y) * z$. A group is a set G on which there is defined an associative law of composition $*$ such that

(*a*) there is an identity element (i.e. an element e such that $(\forall x \in G) e * x = x = x * e$);

(*b*) every element of G has an inverse (i.e. for every $x \in G$ there exists $y \in G$ such that $x * y = e = y * x$).

When the law of composition is written as addition (respectively multiplication) we denote the identity element by 0 (respectively 1) and the inverse of $x \in G$ by $-x$ (respectively x^{-1}). Elements x, y of a group G are said to commute if $xy = yx$, and the group is said to be abelian if every pair of elements commute.

In studying algebraic structures there are two important notions to consider. The first is that of a substructure, and the other is that of a structure-preserving mapping from one such structure to another.

A subgroup of a group G is a non-empty subset H with the property that H is closed (or stable) under the operation of G (i.e. $x, y \in H \Rightarrow xy \in H$) and is also a group under the law of composition that is thus inherited from G. To show that a non-empty subset H is a subgroup of G it suffices to prove that H satisfies the property

$$x, y \in H \Rightarrow xy^{-1} \in H.$$

In additive notation, this becomes

$$x, y \in H \Rightarrow x - y \in H.$$

Alternatively, one can check that H is stable and that $y \in H \Rightarrow y^{-1} \in H$. In order to prove that a particular set H together with a given law of composition forms a group, it is often best to show that H is a subgroup of a larger set which is known to be a group. For example, the set $2\mathbb{Z}$ of even integers is a group under addition, being a subgroup of the additive group $\mathbb{Z}$. If G is a group and S is a non-empty subset of G then the smallest subgroup of G that contains S is denoted by $\langle S \rangle$ and consists of all products of powers of elements of S. In particular, if $S = \{g\}$ then the subgroup $\langle S \rangle = \langle g \rangle$ is said to be cyclic. It is given by:

$$\langle g \rangle = \{g^i \mid i \in \mathbb{Z}\}$$

where $g^0 = 1$. Note that all the powers of g need not be distinct : it is possible to have $g^m = g^n$ with $m \neq n$, so that $g^{m-n} = 1$. The smallest positive integer k (when it exists) such that $g^k = 1$ is called the order of g in G. When no such k exists, G is said to have infinite order. This should not be confused with what is called the order of the group G, namely the number $|G|$ of elements in G. Nevertheless, note that $|\langle g \rangle|$ is equal to the order of g. For every positive integer n there is a cyclic group of order n. This is denoted by C_n. All subgroups of C_n are of the form C_m where m divides n. The celebrated theorem of Lagrange states that for a finite group the order of every subgroup divides the order of the group; i.e. if H is a subgroup of G then $|H|$ is a factor of $|G|$.

The usual proof of Lagrange's theorem uses the notion of a coset. If H is a subgroup of G then the right cosets (respectively left cosets) of H in G are the sets $Hx = \{hx \mid h \in H\}$ (respectively $xH = \{xh \mid h \in H\}$). The set of right (respectively left) cosets of H in G forms a partition of G and the number of cosets is called the index of H in G. A subgroup H of G is said to be normal in G if $(\forall x \in G)xH = Hx$. When H is normal, the set of cosets forms a group under the law of composition given by $xH.yH = xyH$. This group is called the quotient group of G by H and is written G/H.

The notion of a normal subgroup is intimately related to that of a structure-preserving mapping. If G, H are groups then a mapping $f: G \to H$ is called a group morphism if $(\forall x, y \in G) f(xy) = f(x)f(y)$. Group morphisms are very often called homomorphisms. A bijective group morphism is called an isomorphism. There are two important subsets associated with every group morphism $f: G \to H$, namely the kernel and image of f, defined respectively by

$$\operatorname{Ker} f = \{x \in G \mid f(x) = 1\};$$
$$\operatorname{Im} f = \{f(x) \mid x \in G\}.$$

For every group morphism $f: G \to H$, $\operatorname{Ker} f$ is a normal subgroup of G. Conversely, every normal subgroup is the kernel of some group morphism. The first isomorphism theorem for groups states that if $f: G \to H$ is a group morphism then $\operatorname{Im} f$ is isomorphic to $G/\operatorname{Ker} f$; in symbols, $\operatorname{Im} f \cong G/\operatorname{Ker} f$.

We assume that the reader is familiar with the notion of a permutation (simply a bijection on a finite set) and the notation

$$\begin{pmatrix} 1 & 2 & 3 & \dots & n \\ f(1) & f(2) & f(3) & \dots & f(n) \end{pmatrix}$$

as well as the cycle notation for permutations. Here we treat permutations precisely as mappings, so that multiplication of permutations is their composition as mappings. For example,

$$\begin{pmatrix} 1 & 2 & 3 \\ 3 & 2 & 1 \end{pmatrix}\begin{pmatrix} 1 & 2 & 3 \\ 2 & 1 & 3 \end{pmatrix} = \begin{pmatrix} 1 & 2 & 3 \\ 2 & 3 & 1 \end{pmatrix},$$

or, in cycle notation,

$$(1\ 3)(1\ 2) = (1\ 2\ 3).$$

The set of all permutations on a set of n elements (which is usually chosen to be $\{1, 2, \dots, n\}$) is a multiplicative group denoted by S_n.

As an illustration of the first isomorphism theorem, consider the mapping $\vartheta : S_3 \to S_3$ defined by

$$\vartheta(x) = \begin{cases} (1) & \text{if } x \text{ is even;} \\ (1\ 2) & \text{if } x \text{ is odd.} \end{cases}$$

ϑ is a group morphism and $S_2 \cong \text{Im}\ \vartheta \cong S_3/\text{Ker}\ \vartheta$ where $\text{Ker}\ \vartheta = \{(1), (1\ 2\ 3), (3\ 2\ 1)\}$ is the cyclic subgroup of S_3 generated by $\{(1\ 2\ 3)\}$.

1.1 Show that the prescription

$$[a]\,[b] = [ab]$$

defines a binary operation on the set $\mathbb{Z}_n$ of congruence classes modulo n. Show also that the prescription

$$[a] + [b] = [a + b]$$

defines a binary operation on $\mathbb{Z}_n$.

Find $[a]$, $[b] \in \mathbb{Z}_6$ with $[a] \neq [0]$ so that the equation

$$[a]\,[x] = [b]$$

has (*a*) no solution, (*b*) exactly one solution for $[x] \in \mathbb{Z}_6$. Is it possible to find $[a]$, $[b] \in \mathbb{Z}_6$ so that the equation has more than one solution?

1.2 Let R be the equivalence relation defined on $\mathbb{Z}$ by

$$xRy \text{ if and only if } x^2 \equiv y^2 \pmod 6.$$

Let S denote the set of equivalence classes. Show that $[x]_R\,[y]_R = [xy]_R$ defines a binary operation on S and compile the associated Cayley table. Does $[x]_R + [y]_R = [x + y]_R$ define a binary operation on S?

1.3 Let $P = \{p \in \mathbb{Z} \mid p \text{ is a prime and } p \leqslant 13\}$. Define a binary operation $*$ on P by

$$p * q = \text{the greatest prime divisor of } p + q - 2.$$

Construct the Cayley table for $*$ and show that P has an identity with respect to $*$. Does every element of P have an inverse? Is $*$ associative?

1.4 A binary operation $*$ is defined on $E = \mathbb{R} \times \mathbb{R} \times \mathbb{R}$ by

$$(x, y, z) * (x', y', z') = (xx', \alpha y' + yx', \alpha z' + zx')$$

where $\alpha = x + y + z$. Show that $*$ is associative and that E contains an identity with respect to $*$. Show also that (x, y, z) has an inverse if and only if x and α are both non-zero. Find the inverse of (x, y, z) when it exists.

1.5 For each of the following sets, determine whether the given operation is associative. Does the set contain an identity for the given operation? If it has an identity does every element have an inverse? Is the set a group under the given operation?

(*a*) The even integers under addition.
(*b*) The rationals under subtraction.
(*c*) The rationals under multiplication.
(*d*) The positive rationals under multiplication.
(*e*) $\mathbb{Z}_6 \setminus \{[0]\}$ under multiplication.
(*f*) $\mathbb{Z}_5$ under multiplication.
(*g*) The cube roots of 1 under multiplication.

1.6 Which of the following sets are groups under the operation $*$ given?
(*a*) Positive integers with $a * b = \max\{a, b\}$.
(*b*) $\mathbb{Z}$ with $a * b = \min\{a, b\}$.
(*c*) $\mathbb{R}$ with $a * b = a + b - ab$.
(*d*) Positive integers with $a * b = \max\{a, b\} - \min\{a, b\}$.
(*e*) $\mathbb{Z} \times \mathbb{Z}$ with $(a, b) * (c, d) = (a + c, b + d)$.
(*f*) $\mathbb{R} \times \mathbb{R}$ with $(a, b) * (c, d) = (ac + bd, ad + bd)$.
(*g*) $(\mathbb{R} \times \mathbb{R}) \setminus \{(0, 0)\}$ with $(a, b) * (c, d) = (ac - bd, ad + bc)$.

1.7 Let T be the set of matrices of the form

$$\begin{bmatrix} x & y \\ -y & x \end{bmatrix}$$

where $x, y \in \mathbb{R}$. Prove that T is a group under matrix addition, and that $T \setminus \{0\}$ is a group under matrix multiplication.

1.8 Let G be the set of mappings $f: \mathbb{R} \to \mathbb{R}$ of the form $f(x) = ax + b$ with $a \neq 0$. Prove that G forms a group under composition of mappings.

1.9 Define a binary operation $*$ on $\mathbb{R}^* = \mathbb{R} \setminus \{0\}$ by

$$a * b = \begin{cases} ab & \text{if } a > 0; \\ \dfrac{a}{b} & \text{if } a < 0. \end{cases}$$

Is $\mathbb{R}^*$ a group under $*$?

1.10 Determine the group of symmetries of each of the figures in Fig. 1.2.

Fig.1.2

(*a*)
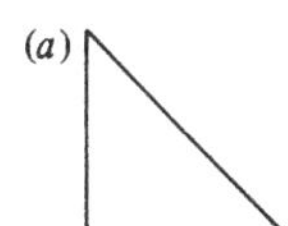
(*b*)
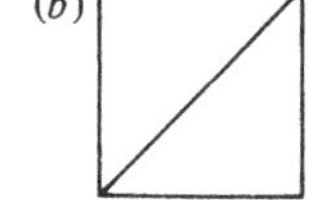
(*c*)
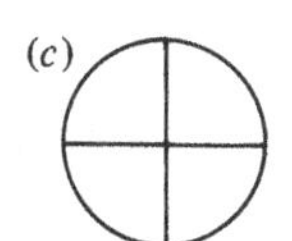
(*d*)
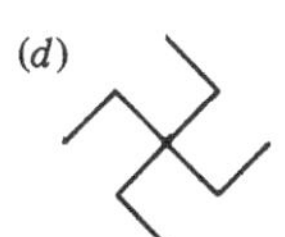

1.11 Prove that the group of symmetries of a regular n-sided polygon (with $n \geqslant 3$) has $2n$ elements and describe them. Prove that none of these *dihedral groups* is abelian.

1.12 Consider the infinite pattern

... ∀∀∀∀∀ ...

Clearly the symmetries of this consist of

(*a*) translations a number of units in either direction;

(*b*) reflections in a vertical line either through the centre of any symbol or between two symbols.

Let a be the translation one unit to the right and let b be the reflection in a vertical line through the centre of a fixed symbol. Show that the group of symmetries is generated by a and b subject to the relations $b^2 = 1, ba = a^{-1}b$.

There are seven 'linear' patterns with different symmetry groups. Can you find them all?

1.13 Let G be the group of rotations of a cube. Show that $|G| = 24$. Find eight different non-identity elements satisfying $x^3 = 1$ which fix a corner. Find nine non-identity elements satisfying $x^4 = 1$ which take a given face to itself. Find six non-identity elements satisfying $x^2 = 1$ which take a given edge to itself. Check that each of the 24 elements has now been identified.

Do the same for the group of rotations of an octahedron (the regular solid with six vertices and eight triangular faces).

(*Hint*: This can be deduced easily from the group of rotations of a cube!)

1.14 The following is part of the Cayley table of a finite group. Fill in the missing entries:

	e	a	b	x	y	z
e	e	a	b	x	y	z
a	a	b	e	y		
b	b					
x	x	z				a
y	y					
z	z					

1.15 Use Lagrange's theorem to show that if G is a group with $|G| = n$ then $g^n = 1$ for every $g \in G$. Deduce that if p is a prime then $a^p \equiv a \pmod p$ for any integer a. Hence find the remainder on dividing

(a) 7^{100} by 11;

(b) 9^{37} by 13.

1.16 Let G be a group and let $a, b \in G$ be such that $ab = ba^k$ where k is a fixed positive integer. Prove by induction that, for all positive integers n,

(a) $a^n b = ba^{kn}$;

(b) $ab^n = b^n a^{k^n}$.

Suppose now that b is of order 2. Prove that $(ba^n)^2$ commutes with both a and b.

1.17 Let G be a group and let $a, b \in G$ be such that $ab \neq ba$. Prove that the elements $1, a, b, ab, ba$ are distinct. Show further that either $a^2 = 1$ or $a^2 \notin \{1, a, b, ab, ba\}$. In the case where $a^2 = 1$ show that $aba \notin \{1, a, b, ab, ba\}$. Deduce that every non-abelian group contains at least six elements.

Let F be the set of six functions from $\mathbb{R} \setminus \{0, 1\}$ to itself given by

$$e(x) = x, \quad p(x) = \frac{1}{1-x}, \quad q(x) = \frac{x-1}{x},$$

$$a(x) = \frac{1}{x}, \quad b(x) = 1-x, \quad c(x) = \frac{x}{x-1}.$$

Construct the Cayley table for F under $\circ$ and show that F is a non-abelian group.

1.18 (a) Let a and b be elements of a group G which commute. Prove that a and b^{-1} commute. Prove also that if x is any element of G then xax^{-1} and xbx^{-1} commute.

(b) Prove that a group G is abelian if and only if $(ab)^{-1} = a^{-1}b^{-1}$ for all $a, b \in G$.

(c) Prove that if G is a group such that $(ab)^2 = a^2b^2$ for all $a, b \in G$ then G is abelian.

(d) Let G be a group and suppose that $a^2 = 1$ for every $a \in G$. Prove that G is abelian.

(e) If G is a group and $x, y \in G$ prove by induction that, for every positive integer n, $(x^{-1}yx)^n = x^{-1}y^n x$.

(f) Let G be a group and suppose that $a, b \in G$ are such that $b^6 = 1$ and $ab = b^4 a$. Prove that $b^3 = 1$ and that $ab = ba$.

1.19 Let G be the multiplicative group of non-singular 2×2 matrices with rational entries. What are the orders of the following elements of G?

$$\begin{bmatrix} 0 & -1 \\ 1 & 1 \end{bmatrix}, \begin{bmatrix} -\frac{1}{2} & \frac{1}{2} \\ -\frac{3}{2} & -\frac{1}{2} \end{bmatrix}, \begin{bmatrix} 0 & -1 \\ 1 & 0 \end{bmatrix}.$$

Find an element of order 2 in G.

Prove by induction that if

$$A = \begin{bmatrix} 1 & 1 \\ 0 & 1 \end{bmatrix} \text{ then } A^n = \begin{bmatrix} 1 & n \\ 0 & 1 \end{bmatrix}$$

for every $n \in \mathbb{N}$. Deduce that A does not have finite order. Is G a finite group? Find $a, b, c \in G$ such that a and b commute, b and c commute but a and c do not commute.

1.20 Which of the following groups are abelian? Which are cyclic? Which are finite?

(*a*) $G = \left\{ \begin{bmatrix} 1 & n \\ 0 & 1 \end{bmatrix} \mid n \in \mathbb{Z} \right\}$ under multiplication.

(*b*) The non-zero elements of $\mathbb{Z}_{11}$ under multiplication.

(*c*) The group of bijections of $E = \{1, 2, 3\}$ to itself under composition of mappings.

(*d*) The group of bijections of $E = \{1, 2\}$ to itself under composition of mappings.

(*e*) The even integers under addition.

(*f*) $\mathbb{R}$ under addition.

1.21 On the set $G = \mathbb{R} \setminus \{0\} \times \mathbb{R}$ define a law of composition by

$$(a, b)(c, d) = (ac, bc + d).$$

Prove that G is a non-abelian group.

Determine which of the following subsets of G are subgroups.

$H = \{(a, k(a-1)) \mid a \neq 0\}$ where $k \in \mathbb{R}$ is fixed;

$K = \{(a, 0) \mid a > 0\}$;

$L = \{(a, na^n) \mid a \neq 0\}$ where $n \in \mathbb{N}$ is fixed;

$M = \{(1, b) \mid b \in \mathbb{R}\}$.

Show that G contains an infinite number of elements of order 2. Does G contain elements of order 3?

1.22 (*a*) Let G denote the group of non-zero rationals under multiplication. Let

$$H = \{2^n \mid n \in \mathbb{Z}\}, \quad K = \left\{ \frac{1+2n}{1+2m} \mid n, m \in \mathbb{Z} \right\}.$$

Are H and K subgroups of G?

(*b*) Show that $H = \{[0], [4], [8], [12]\}$ is a subgroup of $\mathbb{Z}_{16}$ under addition.

(*c*) Find all the subgroups of the group in question 1.14.

1.23 Let $S = \{(x, y) \mid x, y \in \mathbb{R}, x \neq 0, x + y \neq 0\}$. Prove that the binary operation defined on S by

$$(x_1, y_1)(x_2, y_2) = (x_1x_2, (x_1 + y_1)(x_2 + y_2) - x_1x_2)$$

makes S into a group.

Show that $P = \{(1, y) \mid y > -1\}$ is a subgroup of S.

1.24 Prove that

$$G = \left\{ \begin{bmatrix} a & b \\ c & d \end{bmatrix} \mid a, b, c, d \in \mathbb{Z}, ad - bc = 1 \right\}$$

is a group under multiplication. Let H_N be the subset of G consisting of those matrices $\begin{bmatrix} a & b \\ c & d \end{bmatrix}$ in G with

$$a \equiv d \equiv 1 (\text{mod } N), \quad b \equiv c \equiv 0 (\text{mod } N),$$

N being a fixed positive integer. Prove that H_N is a subgroup of G.

If $S = \{H_N \mid 2 \leqslant N \leqslant 12\}$ is ordered by a set inclusion, draw the Hasse diagram for S.

1.25 Let p be an odd prime. A non-zero integer a is said to be a *quadratic residue* of p if there exists an integer x such that $x^2 \equiv a (\text{mod } p)$. Prove that

(*a*) the quadratic residues of p form a subgroup Q of the group of non-zero integers mod p under multiplication;

(*b*) $|Q| = \frac{1}{2}(p - 1)$;

(*c*) if $q \in Q, n \notin Q$ then nq is not a quadratic residue;

(*d*) if m and n are not quadratic residues then mn is a quadratic residue;

(*e*) if a is a quadratic residue then $a^{(p-1)/2} \equiv 1 (\text{mod } p)$.

1.26 Let $M = \{M(\vartheta) \mid \vartheta \in \mathbb{R}\}$ where

$$M(\vartheta) = \begin{bmatrix} \cos \vartheta & -\sin \vartheta \\ \sin \vartheta & \cos \vartheta \end{bmatrix}.$$

Prove that M is an abelian group. Show that M contains cyclic subgroups of every finite order. Does M contain infinite cyclic subgroups?

1.27 Let $G = \langle g \rangle$ be a cyclic group of finite order n generated by the element

g. What are the subgroups of G? Prove that the other generators of G are the elements g^r where hcf $\{r, n\} = 1$.

For the group $\mathbb{Z}_{18}$ list all the subgroups and find all the generators in each subgroup. Draw the subgroup Hasse diagram.

1.28 Prove that if G is a group and H, K are subgroups of G then $H \cap K$ is a subgroup of G. Suppose now that G is finite and that hcf $\{|H|, |K|\} = 1$. Prove that $H \cap K = \{1\}$.

Find $m\mathbb{Z} \cap n\mathbb{Z}$ in the additive group $\mathbb{Z}$. Deduce that no two non-trivial subgroups of $\mathbb{Z}$ can intersect in the trivial subgroup.

1.29 Suppose that G is a group. For every $a \in G$ define

$$C(a) = \{x \in G \mid ax = xa\}.$$

Prove that $C(a)$ is a subgroup of G.

Now suppose that H is a subgroup of G. Define

$$C(H) = \{x \in G \mid (\forall a \in H) ax = xa\}.$$

Prove that $C(H)$ is a subgroup of G.

Give examples of situations where

(*a*) $C(H) = H$;

(*b*) $C(H) = \{1\}$;

(*c*) $H \neq G$ and $C(H) = G$.

1.30 If G is an abelian group and $a, b \in G$ are distinct elements of order 2, show that ab has order 2. Prove that $\{1, a, b, ab\}$ forms a subgroup of G that is not cyclic.

Consider the set

$$G = \{[n] \mid n \text{ an odd integer}\},$$

where $[n]$ denotes the congruence class of n modulo 2^{k+1} and k is a fixed integer greater than 1. Given that G is a group under the usual multiplication of congruence classes show, by considering the elements $[2^k - 1]$ and $[2^k + 1]$, that G is not cyclic.

1.31 Let $\mathbb{Q}$ be the additive group of rationals. If

$$r_1 = \frac{p_1}{q_1} \in \mathbb{Q} \text{ and } r_2 = \frac{p_2}{q_2} \in \mathbb{Q}$$

prove that

$$\langle r_1, r_2 \rangle \subseteq \left\langle \frac{1}{q_1 q_2} \right\rangle.$$

Extend this result to any finite collection of elements of $\mathbb{Q}$ and deduce that if $r_1, \ldots, r_n \in \mathbb{Q}$ then $\langle r_1, \ldots, r_n \rangle$ is cyclic.

Show that if H_n is the subgroup of $\mathbb{Q}$ generated by $1/n!$ then

$$H_1 \subseteq H_2 \subseteq H_3 \subseteq \cdots$$

with each H_n cyclic. Show, moreover, that each of these inclusions is strict. Prove that $\mathbb{Q} = \bigcup_{n \geqslant 1} H_n$, so that $\mathbb{Q}$ is a union of cyclic groups.

Prove finally that $\mathbb{Q}$ is not cyclic.

1.32 Let G be a group and let H be a subgroup of G. Prove that the following conditions on H are equivalent:

(*a*) $(\forall x \in G)xH = Hx$ (i.e. H is normal);
(*b*) $(\forall x \in G)x^{-1}Hx \subseteq H$;
(*c*) $(\forall x \in G)x^{-1}Hx = H$;
(*d*) $(\forall x \in G)(\forall h \in H)x^{-1}hx \in H$;
(*e*) $(\forall x \in G)(\forall h \in H)(\exists h' \in H)xh = h'x$;
(*f*) $(\forall x \in G)(\forall h \in H)h^{-1}x^{-1}hx \in H$.

1.33 Let N be a subgroup of a group G. Prove that the following conditions are equivalent:

(1) N is normal in G;
(2) $(\forall x \in G)(\forall n \in N)xN = nxN$.

Are the following conditions equivalent?

(*a*) N is normal in G;
(*b*) $(\forall x \in G)(\forall n \in N)xnx \in N$.

1.34 If a subgroup H has index 2 in a group G prove that H is normal in G.

Let A be the set of matrices of the form $\begin{bmatrix} a & 0 \\ 0 & b \end{bmatrix}$ and let B be the set of matrices of the form $\begin{bmatrix} 0 & a \\ b & 0 \end{bmatrix}$ where a, b are non-zero real numbers. Prove that $G = A \cup B$ is a group under matrix multiplication. Prove also that A is a subgroup of G and that A, B are the cosets of A in G. Deduce that A is normal in G.

Show that the set of matrices of the form $\begin{bmatrix} a & 0 \\ 0 & 1 \end{bmatrix}$ is a normal subgroup of A. Deduce that 'a normal subgroup of a normal subgroup is not necessarily normal'.

1.35 Consider the set G of matrices of the form

$$\begin{bmatrix} 1 & a & b \\ 0 & 1 & c \\ 0 & 0 & 1 \end{bmatrix}$$

where a, b, $c \in \mathbb{Z}$. Show that G is a group under matrix multiplication. Prove that the subset H of matrices of the form

$$\begin{bmatrix} 1 & 0 & a \\ 0 & 1 & 0 \\ 0 & 0 & 1 \end{bmatrix},$$

where $a \in \mathbb{Z}$, is a normal subgroup of G. Prove also that the set K of matrices of the form

$$\begin{bmatrix} 1 & a & 0 \\ 0 & 1 & 0 \\ 0 & 0 & 1 \end{bmatrix},$$

where $a \in \mathbb{Z}$, is a subgroup of G. Is K a normal subgroup?

1.36 Let G be a group with the property that there exists a positive integer n such that $(\forall x, y \in G)(xy)^n = x^n y^n$. Prove that the subsets

$$G^n = \{x^n \mid x \in G\} \quad \text{and} \quad G_n = \{x \in G \mid x^n = 1\}$$

are normal subgroups of G and that $G^n \cong G/G_n$.

1.37 (*a*) Write the element

$$\begin{pmatrix} 1 & 2 & 3 & 4 & 5 & 6 & 7 & 8 & 9 \\ 4 & 6 & 9 & 7 & 2 & 5 & 8 & 1 & 3 \end{pmatrix}$$

of S_9 as a product of disjoint cycles.

(*b*) Express each of the following permutations as products of disjoint cycles:

$$\begin{pmatrix} 1 & 2 & 3 & 4 & 5 & 6 & 7 & 8 \\ 8 & 2 & 6 & 3 & 7 & 4 & 5 & 1 \end{pmatrix}, \quad \begin{pmatrix} 1 & 2 & 3 & 4 & 5 & 6 & 7 & 8 \\ 3 & 6 & 4 & 1 & 8 & 2 & 5 & 7 \end{pmatrix},$$

$$\begin{pmatrix} 1 & 2 & 3 & 4 & 5 & 6 & 7 & 8 \\ 3 & 1 & 4 & 7 & 2 & 5 & 8 & 6 \end{pmatrix}.$$

(*c*) Write $(143)(534)(137)$ as a product of disjoint cycles.

1.38 Do the elements $(12)(34)$ and $(13)(24)$ of S_4 commute? Do the elements $(12)(24)$ and $(13)(34)$ of S_4 commute?

1.39 Prove that the order of a cycle of length r in S_n is r.

If a permutation σ is a product of disjoint cycles of lengths r and s, prove that the order of σ is the lowest common multiple of r and s.

Express the permutation

$$\sigma = \begin{pmatrix} 1 & 2 & 3 & 4 & 5 & 6 & 7 & 8 & 9 \\ 3 & 7 & 8 & 9 & 4 & 5 & 2 & 1 & 6 \end{pmatrix}$$

as a product of disjoint cycles. Hence determine the order of σ. Compute σ^{1000}.

1.40 (*a*) Find $a^{-1}ba$ when $a = (135)(12)$ and $b = (1579)$. What is the order of $a^{-1}ba$? Do the same for $a = (579)$ and $b = (123)$.

(*b*) Compute $\tau\sigma\tau^{-1}$ where $\tau = (135)(24)$ and $\sigma = (12)(3456)$.

1.41 If σ is a permutation of $1, 2, \ldots, n$ and if $(k_1 k_2 \ldots k_r)$ is a cycle, prove that

$$\sigma(k_1 k_2 \ldots k_r)\sigma^{-1} = (\sigma(k_1)\sigma(k_2) \ldots \sigma(k_r)).$$

Deduce that $\{(1), (12)(34), (13)(24), (14)(23)\}$ is a normal subgroup of S_4.

1.42 Let G_1, G_2 be groups and define a binary operation on $G_1 \times G_2$ by

$$(g_1, g_2)(g_1', g_2') = (g_1 g_1', g_2 g_2').$$

Show that this makes $G_1 \times G_2$ into a group. Prove that $H_1 = G_1 \times \{1\}$ and $H_2 = \{1\} \times G_2$ are normal subgroups of the cartesian product group $G_1 \times G_2$. Show that if $h_1 \in H_1$ and $h_2 \in H_2$ then h_1 and h_2 commute.

1.43 Which of the following are group morphisms? Find the image and kernel of those that are morphisms.

(*a*) $f: \mathbb{Z}_{12} \to \mathbb{Z}_{12}$, $f([x]_{12}) = [x+1]_{12}$;
(*b*) $f: C_{12} \to C_{12}$, $f(g) = g^3$;
(*c*) $f: \mathbb{Z} \to \mathbb{Z}_2 \times \mathbb{Z}_4$, $f(x) = ([x]_2, [x]_4)$;
(*d*) $f: \mathbb{Z}_8 \to \mathbb{Z}_2$, $f([x]_8) = [x]_2$;
(*e*) $f: C_2 \times C_3 \to S_3$, $f(h^r, k^s) = (12)^r(123)^s$;
(*f*) $f: S_n \to S_{n+1}$, $f(\pi)$ is the permutation given by $i \to \pi(i)$ for $i \leqslant n$ and $n+1 \to n+1$.

1.44 Find all the morphisms from the additive group $\mathbb{Z}$ to the additive group $\mathbb{Q}$.

1.45 Prove that the set G of matrices of the form,

$$\begin{bmatrix} 1-n & -n \\ n & 1+n \end{bmatrix},$$

where $n \in \mathbb{Z}$, is an abelian group under multiplication. Show that $G \cong \mathbb{Z}$.

Show also that the set G_1 of matrices of the form

$$\begin{bmatrix} 1-2n & n \\ -4n & 1+2n \end{bmatrix},$$

where $n \in \mathbb{Z}$, is a group. Is $G_1 \cong G$?

1.46 Let $\mathbb{Z}_2$ be the integers modulo 2 and let G be the group of matrices of the form

$$\begin{bmatrix} a & b \\ c & d \end{bmatrix} a, b, c, d \in \mathbb{Z}_2, ad - bc = 1$$

where the operation is usual matrix multiplication with the entries reduced modulo 2. If T is the set consisting of the matrices

$$\begin{bmatrix} 0 \\ 1 \end{bmatrix}, \begin{bmatrix} 1 \\ 0 \end{bmatrix}, \begin{bmatrix} 1 \\ 1 \end{bmatrix}$$

and if

$$A = \begin{bmatrix} 0 & 1 \\ 1 & 0 \end{bmatrix}$$

show that $X \to AX$ describes a permutation on T. Hence show that $G \cong S_3$.

1.47 Show that the group $\mathbb{Z}[X]$ under addition is isomorphic to the group $\mathbb{Q}^+$ of positive rationals under multiplication.

(*Hint*: Consider $\vartheta : \mathbb{Z}[X] \to \mathbb{Q}^+$ given by

$$a_0 + a_1X + \cdots + a_nX^n \to p_0^{a_0}p_1^{a_1} \ldots p_n^{a_n}$$

where $p_0 < p_1 < \cdots < p_n$ are the first $n+1$ primes.)

1.48 Let ϑ be a fixed real number and let

$$A = \begin{bmatrix} 0 & 1 & -\sin\vartheta \\ -1 & 0 & \cos\vartheta \\ -\sin\vartheta & \cos\vartheta & 0 \end{bmatrix}.$$

Show that $A^3 = 0$.

Given $x \in \mathbb{R}$, define the 3×3 matrix A_x by

$$A_x = I_3 + xA + \tfrac{1}{2}x^2A^2$$

where I_3 is the 3×3 identity matrix. Prove that, under matrix multiplication, $G = \{A_x \mid x \in \mathbb{R}\}$ is an abelian group which is isomorphic to the additive group $\mathbb{R}$.

1.49 Show that $G = \{(a, b) \in \mathbb{R} \times \mathbb{R} \mid b \neq 0\}$ forms a group under the operation defined by

$$(a, b)(c, d) = (a + bc, bd).$$

Show that G is not abelian. Show also that the subsets

$$H = \{(a, b) \in G \mid a = 0\},$$
$$K = \{(a, b) \in G \mid b > 0\},$$
$$L = \{(a, b) \in G \mid b = 1\}$$

are subgroups of G. Which of these subgroups are abelian?

Show that $H \cap K$ is a group isomorphic to the multiplicative group $\mathbb{R}^+$ of positive reals.

Find all the elements of order 2 in G.

1.50 On the set $G = \mathbb{Z} \times \mathbb{Z}$ define a binary operation by

$$(a, b)(c, d) = (a + (-1)^b c, b + d).$$

Prove that G is a non-abelian group.

Let $H = \{(a, b) \in G \mid b = 0\}$ and $K = \{(a, b) \in G \mid a = 0\}$. Prove that H and K are subgroups of G and determine whether or not they are normal.

If $f : G \to G$ is given by $f(a, b) = (0, b)$ prove that f is a group morphism and find its image and kernel. Hence prove that the groups G/H and K are isomorphic.

1.51 Let A be an abelian group and let $E = \{1, -1\}$. Define a binary operation on $G = A \times E$ by

$$(a, \epsilon)(b, \delta) = (ab^{\epsilon}, \epsilon\delta).$$

Prove that G is a group. Give examples where G is (*a*) abelian, (*b*) non-abelian.

Show that $N = \{(a, 1) \mid a \in A\}$ is a normal subgroup of G. Prove that $N \cong A$ and identify the group G/N.

Show that, with a suitable choice for the group A, the group G is isomorphic to the group in question 1.9.

1.52 Let G be an abelian group and let $x, y \in G$. Prove that $(xy)^n = x^n y^n$ for all positive integers n, and deduce that the subset H consisting of all the elements of finite order is a subgroup of G.

Suppose now that G is the multiplicative group $\{z \in \mathbb{C} \mid |z| = 1\}$. Prove that in this case $H = \{e^{\alpha \pi i} \mid \alpha \in \mathbb{Q}\}$.

If $e^{\vartheta i} = e^{\varphi i}$ show that

$$\frac{\vartheta}{\pi} - \frac{\varphi}{\pi} \in \mathbb{Q}.$$

Deduce that the prescription

$$f(e^{\vartheta i}) = \frac{\vartheta}{\pi} + \mathbb{Q}$$

defines a mapping $f : G \to \mathbb{R}/\mathbb{Q}$. Prove that f is a group morphism and hence show that $G/H \cong \mathbb{R}/\mathbb{Q}$.

1.53 Let G be the subgroup of the symmetric group S_4 generated by the permutations (12)(34) and (123). Show that the subgroup H of G generated by (12)(34) and (13)(24) is a normal subgroup such that G/H is the cyclic group of order 3.

Show that the subgroup K of H generated by (12)(34) is normal in H but not normal in G.

1.54 Let G be the subgroup of the symmetric group S_5 generated by (12345) and (25)(34). Show that the subgroup H of G generated by (12345) is a normal subgroup such that G/H is cyclic. Find the order of G/H and hence determine the order of G. Show that G is isomorphic to the symmetry group of the regular pentagon.

1.55 Let m and n be integers with m a factor of n. Show that $n\mathbb{Z}$ is a subgroup of $m\mathbb{Z}$ and that $m\mathbb{Z}/n\mathbb{Z}$ is a subgroup of $\mathbb{Z}/n\mathbb{Z}$. Show that the prescription

$$f(a + n\mathbb{Z}) = a + m\mathbb{Z}$$

defines a mapping $\mathbb{Z}/n\mathbb{Z} \to \mathbb{Z}/m\mathbb{Z}$. Show also that f is a group morphism and find its kernel. Hence show that

$$\frac{\mathbb{Z}/n\mathbb{Z}}{m\mathbb{Z}/n\mathbb{Z}} \cong \frac{\mathbb{Z}}{m\mathbb{Z}}.$$

1.56 Let X be a set. Show that $\mathbf{P}(X)$ is a group under the binary operation Δ of symmetric difference. Prove that this group is abelian and that every element has order 2.

If $X = \{1, 2\}$ show that $\mathbf{P}(X) \cong C_2 \times C_2$.

1.57 Let H be a subgroup of a finite group G. Let S be the set of left cosets of H in G. If $g \in G$ show that the mapping $\varphi_g : S \to S$ given by

$$\varphi_g(aH) = gaH$$

is a bijection. Use this to define a group morphism $\vartheta : G \to S_k$ where $k = |S|$. Prove that Ker $\vartheta \subseteq H$. Show, moreover, that Ker ϑ is the largest normal subgroup of G which is a subgroup of H. Prove that if $|G|$ does not divide $k!$ then H contains a non-trivial normal subgroup of G.

Finally, show that if H is a subgroup of order 11 in a group G of order 99 then H is normal in G.

1.58 Each of the following groups has order 8:

(*a*) $\mathbf{P}(X)$ under Δ where $X = \{1, 2, 3\}$;
(*b*) $\mathbb{Z}_8$;
(*c*) the group of symmetries of a square;
(*d*) $\{1, 2, 4, 7, 8, 11, 13, 14\}$ under multiplication modulo 15.

Show that no two are isomorphic.

1.59 Show that the set of isomorphisms from a group G to itself forms a group under composition of mappings (the *automorphism group* Aut G of G). Prove that

(*a*) $\text{Aut}\ \mathbb{Z} \cong C_2$;
(*b*) $\text{Aut}\ \mathbb{Z}_n$ is isomorphic to the multiplicative group of integers that are coprime to n;
(*c*) $\text{Aut}\,(\mathbb{Z}_2 \times \mathbb{Z}_2) \cong S_3$.

1.60 Let Q be the quaternion group $\{\pm 1, \pm \mathrm{i}, \pm \mathrm{j}, \pm \mathrm{k}\}$ of order 8. Let $H = Q \times Q$, let $Z^* = \{(1, 1), (-1, -1)\}$ and let $G = H/Z^*$.

Prove that the centre $Z(G)$ of G has order 2. Show also that $G/Z(G)$ is abelian. Deduce that it is possible to write G as a cartesian product group $G_1 \times G_2$ only if G_1 or G_2 is the trivial group.

1.61 Let S be the set of real 2×2 matrices X such that $X + I_2$ is invertible. Show that if $A, B \in S$ then $A + B + AB \in S$.

Prove that S is a group under the law of composition

$$A * B = A + B + AB.$$

If M is the group of all invertible real 2×2 matrices, prove that the groups S and M are isomorphic.

1.62 Let $G = \{x \in \mathbb{R} \mid x^2 < 1\}$. Show that if $x, y \in G$ then

$$\frac{x + y}{1 + xy} \in G.$$

Prove that G is a group under the law of composition

$$x * y = \frac{x + y}{1 + xy}.$$

By considering the mapping $f : G \to \mathbb{R}$ described by

$$f(x) = \log \frac{1+x}{1-x},$$

prove that the group G and the additive group $\mathbb{R}$ are isomorphic.

1.63 Let G be a group and let $f : G \to H$ be a bijection from G to a set H. Prove that there is one and only one law of compositition $*$ on H such that H is a group with f an isomorphism.

2: Rings and fields

In this section we shall deal with algebraic structures that have two laws of composition. A ring R is a non-empty set with two laws of composition, written as addition and multiplication, such that

(1) R is an abelian group under addition;

(2) the multiplication is associative;

(3) these laws are linked by the distributive laws

$$(\forall x, y, z \in R) \quad (x+y)z = xz + yz \quad \text{and} \quad z(x+y) = zx + zy.$$

When R has a multiplicative identity we often say that R is a ring with a 1, or a ring with an identity. When the multiplication is commutative we say that R is a commutative ring. The additive identity element 0 is called the zero of the ring. In a ring it is possible to have $xy = 0$ with $x \neq 0$ and $y \neq 0$. Such elements are called zero divisors. A commutative ring with a 1 which has no zero divisors is called an integral domain. If the ring R has a 1 and $x \in R$ is such that there exists $y \in R$ with $xy = 1 = yx$ then we say that y is an (hence the) inverse of x in R. If every non-zero element of the commutative ring R with a 1 has an inverse then we say that R is a field. Thus, R is a field if

(*a*) R is an additive abelian group;

(*b*) $R \setminus \{0\}$ is a multiplicative abelian group;

(*c*) the distributive laws hold.

The non-commutative ring $\text{Mat}_{n \times n}(\mathbb{R})$ of $n \times n$ matrices over $\mathbb{R}$ has zero divisors (for $n > 1$). The number systems $\mathbb{Q}$, $\mathbb{R}$, $\mathbb{C}$ form fields. The number system $\mathbb{Z}$ is an integral domain that is not a field. The ring $\mathbb{Z}_n$ of integers modulo n is an integral domain (in fact a field) if and only if n is prime.

As in the case of a group, the notions of substructure and structure-preserving mapping are important. A non-empty subset S of a ring R is closed (or stable) under the laws of R if

$$x, y \in S \Rightarrow x + y \in S \quad \text{and} \quad xy \in S,$$

and is a subring of R if it is a ring under the laws of composition thus inherited from R. To show that S is a subring it suffices to prove that

$$x, y \in S \Rightarrow x - y \in S \quad \text{and} \quad xy \in S.$$

If F is a field then a subfield of F is a subring of the ring F which is also a field. Note that a field can contain subrings that are not fields; e.g. $\mathbb{R}$ contains the subfield $\mathbb{Q}$ and the subring $\mathbb{Z}$.

A left (respectively right) ideal of a ring R is a subring I such that $(\forall r \in R) rI \subseteq I$ (respectively $(\forall r \in R) Ir \subseteq I$) where $rI = \{ri \mid i \in I\}$. By an ideal of R we mean a subring that is both a left ideal and a right ideal. Ideals are to rings what normal subgroups are to groups. Thus for example we can form the quotient ring R/I where I is an ideal. This consists of the quotient group R/I under addition, together with the multiplication

$$(x + I)(y + I) = xy + I.$$

For a subset E of a ring R we denote by (E) the ideal generated by E, i.e. the smallest ideal of R that contains E.

If R, S are rings then a mapping $f : R \to S$ is a ring morphism if

$$(\forall x, y \in R) \quad f(x + y) = f(x) + f(y) \quad \text{and} \quad f(xy) = f(x)f(y).$$

A bijective ring morphism is called a ring isomorphism. As with group morphisms there are the important subsets $\text{Ker } f = \{x \in R \mid f(x) = 0\}$ and $\text{Im } f = \{f(x) \mid x \in R\}$; the former is an ideal of R and the latter is a subring of S. The first isomorphism theorem for rings states that for a ring morphism $f : R \to S$ we have a ring isomorphism $\text{Im } f \cong R/\text{Ker } f$.

We shall also consider the ring $R[X]$ of polynomials with coefficients in a given ring R. In the case where R is a field F, we assume that the reader is acquainted with the division algorithm and the euclidean division of one polynomial by another ($a(X) = b(X)q(X) + r(X)$ where $r = 0$ or $\deg r < \deg b$). Likewise, we assume that the reader can compute the highest common factor of two given polynomials. An irreducible polynomial is one that has no proper polynomial factors. For an irreducible polynomial

$$f(X) = a_0 + a_1 X + a_2 X^2 + \cdots + a_{n-1} X^{n-1} + X^n \in F[X]$$

the quotient ring $F[X]/(f(X))$ is a field which can be regarded as

$$\{b_0 + b_1 x + b_2 x^2 + \cdots + b_{n-1} x^{n-1} \mid b_i \in F\}$$

where x satisfies

$$x^n = -a_{n-1}x^{n-1} - \cdots - a_1 x - a_0.$$

For example, the polynomial $X^2 + 1$ is irreducible over $\mathbb{R}$ and so $\mathbb{R}[X]/(X^2+1)$ is a field that is isomorphic to the field

$$\{b_0 + b_1 x \mid b_i \in \mathbb{R}\}$$

where $x^2 = -1$, i.e. $\mathbb{R}[X]/(X^2+1) \cong \mathbb{C}$.

2.1 (*a*) Let S be the set of mappings $f: \mathbb{R} \to \mathbb{R}$. Define $+$ and $\cdot$ on S by the prescriptions

$$(\forall a \in \mathbb{R}) \quad (f+g)(a) = f(a) + g(a), \quad (f \cdot g)(a) = f(a)g(a).$$

Prove that S becomes a commutative ring with an identity. Is S a ring when multiplication is composition of mappings?

(*b*) Let $S = \mathbb{R} \times \mathbb{R}$ and define $+$ and $\cdot$ on S by

$$(a, b) + (c, d) = (a + c, b + d)$$
$$(a, b) \cdot (c, d) = (ac, bd).$$

Prove that S is a commutative ring with an identity. Find which elements of S have a multiplicative inverse.

2.2 (*a*) Let λ be a fixed real number. Prove that the set $R(\lambda)$ of matrices of the form

$$\begin{bmatrix} x+y & y \\ \lambda y & x \end{bmatrix},$$

where $x, y \in \mathbb{R}$, is a subring of the ring $M = \mathrm{Mat}_{2\times 2}(\mathbb{R})$. Show that $R(\lambda)$ has an identity for every λ. Show also that every non-zero element of the subring $R(-1)$ has a multiplicative inverse. Is the same true of the subring $R(0)$?

(*b*) Prove that the set F of matrices of the form

$$\begin{bmatrix} x & x \\ x & x \end{bmatrix},$$

where $x \in \mathbb{R}$, is a subring of M. Show that F has an identity element that is different from the identity element of M. Prove that every non-zero element of F has a multiplicative inverse. How is this possible when the matrices in F all have zero determinant?

2.3 Prove that in the ring $\mathbb{Z}_n$ the set of non-invertible elements consists of zero and the zero divisors.

2.4 Show that the set K of matrices of the form

$$\begin{bmatrix} a & -b \\ b & a \end{bmatrix}$$

where $a, b \in \mathbb{Z}_3$ forms a subring of $\text{Mat}_{2\times 2}(\mathbb{Z}_3)$. Prove that K is a field. Show that the non-zero elements of K form a cyclic group of order 8 under multiplication. Find a generator of this cyclic group.

2.5 Determine how many solutions the equation $x(x+1) = 0$ has in each of the following rings:

(*a*) $\mathbb{Z}$, (*b*) $\mathbb{Z}_6$, (*c*) $\mathbb{Z}_{41}$, (*d*) $\mathbb{R}$.

2.6 In the ring $\mathbb{Z}_{34}$ find

(*a*) an element a such that every element of $\mathbb{Z}_{34}$ with a multiplicative inverse is a power of a;

(*b*) all elements x with $x^2 = x$.

2.7 Let R be a cummutative ring. Define binary operations $\oplus$ and $\otimes$ on $E = R \times R$ by

$$(a, b) \oplus (c, d) = (a + c, b + d);$$
$$(a, b) \otimes (c, d) = (ac + bd, bc + ad).$$

Show that E is a ring with zero divisors. Prove that if R is an integral domain then the zero divisors in E are of the form (a, a) and $(a, -a)$.

2.8 If D is an integral domain prove that the equation $x^2 = 1$ has only two solutions in D. Use this fact to prove *Wilson's theorem* that if p is prime then $(p-1)! \equiv -1 \pmod{p}$.

2.9 Let E be a ring with identity e. Define an operation $*$ on E by setting

$$(\forall a, b \in E) \quad a * b = a + b - ab.$$

Let G be the subset of E consisting of those elements a for which there exists an element w in E with

$$a * w = w * a = 0.$$

Prove that G is closed under $*$ and that it forms a group under this operation. Show also that if $a \in E$ then a belongs to G if and only if $e - a$ is invertible in E.

2.10 Show that the set H of matrices of the form

$$\begin{bmatrix} a + bi & c + di \\ -c + di & a - bi \end{bmatrix}$$

where $a, b, c, d \in \mathbb{R}$ is a subring of $\mathrm{Mat}_{2\times 2}(\mathbb{C})$. Prove that every non-zero element of H has an inverse. Deduce that H contains no zero divisors. Is H a field?

Let K be the subring of H consisting of those matrices of the form

$$\begin{bmatrix} a+bi & c+di \\ -c+di & a-bi \end{bmatrix}$$

where $a, b, c, d \in \mathbb{Z}$. Find all $X \in K$ with $X^2 = -I_2$.

2.11 Consider the subring $\mathbb{Z}[\mathrm{i}] = \{a + \mathrm{i}b \mid a, b \in \mathbb{Z}\}$ of $\mathbb{C}$. For $a + \mathrm{i}b \in \mathbb{Z}[\mathrm{i}]$ define $f(a + \mathrm{i}b) = a^2 + b^2$. Show that if $\alpha, \beta \in \mathbb{Z}[\mathrm{i}]$ then $f(\alpha\beta) = f(\alpha)f(\beta)$. Deduce that $\alpha \in \mathbb{Z}[\mathrm{i}]$ has a multiplicative inverse if and only if $f(\alpha) = 1$. Hence show that $1, -1, \mathrm{i}, -\mathrm{i}$ are the only elements of $\mathbb{Z}[\mathrm{i}]$ that have multiplicative inverses.

If $\alpha \in \mathbb{Z}[\mathrm{i}]$ is non-zero and does not have a multiplicative inverse then we say that α is *irreducible* if $\alpha = \beta\gamma$ requires β or γ to have a multiplicative inverse. Prove that if $f(\alpha)$ is prime then α is irreducible. Give an example of an irreducible element α of $\mathbb{Z}[\mathrm{i}]$ with $f(\alpha)$ not prime.

Write $43\mathrm{i} - 19$ as a product of irreducibles.

2.12 Let R be a commutative ring with an identity. A subset S of R is said to be *multiplicatively closed* if

(1) $0 \notin S$;

(2) $s, t \in S \Rightarrow st \in S$.

Let $M = R \times S$ and define a relation $\sim$ on M by

$$(r_1, s_1) \sim (r_2, s_2) \Leftrightarrow (\exists t \in S)(r_1 s_2 - r_2 s_1)t = 0.$$

(Note that if R has no zero divisors then this is simply $r_1 s_2 = r_2 s_1$.)

Show that $\sim$ is an equivalence relation. If R_S is the set of $\sim$-classes show that addition and multiplication may be defined on R_S by the prescriptions

$$[(r_1, s_1)] + [(r_2, s_2)] = [(r_1 s_2 + r_2 s_1, s_1 s_2)];$$
$$[(r_1, s_1)] \cdot [(r_2, s_2)] = [(r_1 r_2, s_1 s_2)].$$

Show that in this way R_S becomes a commutative ring with an identity.

(*a*) If $R = \mathbb{Z}$ and $S = \mathbb{Z} \setminus \{0\}$ show that $R_S \cong \mathbb{Q}$.

(*b*) If R is an integral domain and $S = R \setminus \{0\}$ show that R_S is a field.

(*c*) If $R = \mathbb{Z}_6$ (which is not an integral domain) and $S = \{2, 4\}$ show that the order of R_S is 3.

2.13 If $p, q \in \mathbb{Q}$ prove that $p^2 - pq + q^2 = 0$ if and only if $p = q = 0$. Use this fact to prove that if

$$I = \begin{bmatrix} 1 & 0 \\ 0 & 1 \end{bmatrix} \quad \text{and} \quad A = \begin{bmatrix} 0 & 1 \\ -1 & -1 \end{bmatrix}$$

then, under matrix addition and multiplication, $S = \{pA + qI \mid p, q \in \mathbb{Q}\}$ is a field.

2.14 Let F be a subfield of the field $\mathbb{R}$ and let p be a positive integer such that $\sqrt{p} \notin F$. Prove that if $a, b \in F$ then $a + b\sqrt{p} = 0$ if and only if $a = b = 0$. Deduce that if $a, b, c, d \in F$ then $a + b\sqrt{p} = c + d\sqrt{p}$ if and only if $a = c$ and $b = d$.

Prove that $F[\sqrt{p}] = \{a + b\sqrt{p} \mid a, b \in F\}$ is a field.

Suppose now that m and n are positive integers such that $\sqrt{m} \notin \mathbb{Q}$, $\sqrt{n} \notin \mathbb{Q}$ and $\sqrt{(mn)} \notin \mathbb{Q}$. Prove that $\sqrt{m} \notin \mathbb{Q}[\sqrt{n}]$ and deduce that the set S of real numbers of the form

$$a + b\sqrt{m} + c\sqrt{n} + d\sqrt{(mn)} \quad (a, b, c, d \in \mathbb{Q})$$

is a field.

2.15 Let R be the set of differentiable functions $f: \mathbb{R} \to \mathbb{R}$. Prove that R is a commutative ring with identity when addition and multiplication are defined by the prescriptions

$$(f+g)(x) = f(x) + g(x), \quad (fg)(x) = f(x)g(x).$$

If $S = \{f \in R \mid f(0) = 0\}$ prove that S is an ideal of R. If $T = \{f \in R \mid \mathrm{D}f(0) = 0\}$, where $\mathrm{D}f$ is the derivative of f, prove that T is a subring of R which is not an ideal of R, but that $S \cap T$ is an ideal of R.

2.16 Let X be a subset of a ring R. Define

$$A(X) = \{r \in R \mid (\forall x \in X) rx = 0\};$$
$$B(X) = \{r \in R \mid (\forall x \in X) xr = 0\}.$$

Prove that $A(X)$ is a left ideal of R and that $B(X)$ is a right ideal of R. If X is a left ideal, prove that $A(X)$ is a two-sided ideal.

Now let $R = \mathrm{Mat}_{2\times 2}(\mathbb{R})$ and let X be the subset of matrices of the form

$$\begin{bmatrix} x & 0 \\ y & 0 \end{bmatrix}.$$

Prove that X is a left ideal of R and find $A(X)$ and $B(X)$. Is $B(X)$ a two-sided ideal of R?

2.17 Let R be a commutative ring and let I be an ideal of R. Define $\sqrt{I}$ to be the set of those $x \in R$ such that $x^n \in I$ for some positive integer n. Prove that $\sqrt{I}$ is an ideal of R, that $I \subseteq \sqrt{I}$, and that $\sqrt{I} = \sqrt{(\sqrt{I})}$.

If J is also an ideal of R, prove that

(*a*) $I \subseteq J \subseteq \sqrt{I} \Rightarrow \sqrt{J} = \sqrt{I}$;

(*b*) $\sqrt{I} \cap \sqrt{J} = \sqrt{(I \cap J)}$.

2.18 Show that the set R of matrices of the form

$$\begin{bmatrix} a & b\sqrt{5} \\ -b\sqrt{5} & a \end{bmatrix}$$

where $a, b \in \mathbb{Z}$ forms a subring of the ring $\mathrm{Mat}_{2\times 2}(\mathbb{R})$. Show that R is commutative and has an identity.

Show that the subset I of matrices of the form

$$\begin{bmatrix} x & (3y+x)\sqrt{5} \\ -(3y+x)\sqrt{5} & x \end{bmatrix}$$

is an ideal of R. Check that each of the matrices

$$\begin{bmatrix} 1 & \sqrt{5} \\ -\sqrt{5} & 1 \end{bmatrix}, \quad \begin{bmatrix} 3 & 0 \\ 0 & 3 \end{bmatrix}$$

belongs to I. By considering the determinants of these matrices, show that there is no $r \in R$ with $I = (r)$.

2.19 Given subsets A and B of a ring R, define

$$A + B = \{a + b \mid a \in A, b \in B\}.$$

Which of the following are true? Give proofs for those that are true and counter-examples (using, say, 2×2 matrices) to illustrate those that are false.

(*a*) If A and B are subrings then so is $A + B$.

(*b*) If A and B are left ideals then so is $A + B$.

(*c*) If A is a left ideal and B is a right ideal then $A + B$ is a subring.

(*d*) If A is a two-sided ideal and B is a left ideal then $A + B$ is a left ideal.

2.20 Let A be an additive abelian group. Let Hom(A) be the set of group morphisms from A to itself. Define an addition on Hom(A) by $(f, g) \to f + g$ where

$$(\forall x \in A) \quad (f+g)(x) = f(x) + g(x),$$

and let multiplication be composition of mappings. Show that Hom(A) is a ring.

If A is the group $\mathbb{Z}$ show that Hom(A) is isomorphic to the ring $\mathbb{Z}$.

If A is the group $\mathbb{Z}_n$ show that Hom(A) is isomorphic to the ring $\mathbb{Z}_n$.

If A is the group $\mathbb{Z} \times \mathbb{Z}$ show, by considering for example the mappings described by $(x, y) \to (x, 0)$ and $(x, y) \to (x, x)$, that Hom(A) is not commutative.

2.21 Let R and S be non-trivial rings and $f: R \to S$ a surjective ring morphism. Determine (giving proofs or counter-examples) which of the following statements are true.

(*a*) If R is commutative then so is S.
(*b*) If R has an identity then so does S.
(*c*) If R and S have identities then $f(1_R) = 1_S$.
(*d*) If R has zero divisors then so does S.
(*e*) If R is an integral domain then so is S.
(*f*) If R is a field then so is S.

2.22 Let F be the ring in question 2.2(*b*). Prove that $f: \mathbb{R} \to \mathrm{Mat}_{2\times 2}(\mathbb{R})$ given by

$$f(x) = \begin{bmatrix} \frac{1}{2}x & \frac{1}{2}x \\ \frac{1}{2}x & \frac{1}{2}x \end{bmatrix}$$

is an injective ring morphism. Deduce that $F \cong \mathbb{R}$.

Define $g: \mathrm{Mat}_{2\times 2}(\mathbb{R}) \to \mathbb{R}$ by

$$\begin{bmatrix} a & b \\ c & d \end{bmatrix} \to a + d.$$

Is g a ring morphism? Are $f \circ g$ and $g \circ f$ ring morphisms?

2.23 Let R be the set of rationals a/b (in lowest terms) such that b is not a multiple of 3. Prove that R is a subring of $\mathbb{Q}$.

Let I be the subset of R consisting of those elements whose numerators are divisible by 3. Prove that I is an ideal of R.

Prove that R/I is a field.

2.24 A ring R is said to be *boolean* if $a^2 = a$ for every $a \in R$. Prove that if R is boolean then $2a = 0$ for every $a \in R$. Deduce that a boolean ring is commutative.

Let $\mathbf{P}(S)$ be the power set of a non-empty set S. Prove that $\mathbf{P}(S)$ is a ring under the operations Δ and $\cap$. Show that this ring is boolean.

If $E \subseteq S$ prove that $\mathbf{P}(E)$ is an ideal of $\mathbf{P}(S)$. If $f_E: \mathbf{P}(S) \to \mathbf{P}(S)$ is given by $f_E(X) = X \cap E'$ (where E' denotes the complement of E in S) prove that f_E is a ring morphism. Deduce that $\mathbf{P}(S)/\mathbf{P}(E) \cong \mathbf{P}(E')$.

2.25 Consider the mapping $\alpha : \mathbb{R}[X] \to \mathrm{Mat}_{2\times 2}(\mathbb{R})$ given by

$$\alpha(f) = \begin{bmatrix} f(0) & 0 \\ \mathrm{D}f(0) & f(0) \end{bmatrix}$$

where $\mathrm{D}f$ denotes the derivative of $f \in \mathbb{R}[X]$. Prove that α is a ring morphism and determine its image and kernel. Find a polynomial that generates the kernel of α.

2.26 Prove that $R = \{a + b\sqrt{2} \mid a,\ b \in \mathbb{Z}\}$ and $S = \{a + b\sqrt{3} \mid a,\ b \in \mathbb{Z}\}$ are subrings of $\mathbb{C}$. Show that R and S are not isomorphic as rings.

2.27 Prove that the natural ring morphism $\vartheta : \mathbb{Z} \to \mathbb{Z}_4$ given by $\vartheta(n) = n \pmod 4$ maps all odd squares to the same element.

Deduce that no member of the sequence of integers

$$11, \quad 111, \quad 1111, \quad 11111, \quad \ldots$$

is a square.

2.28 Prove that if p is prime then there are exactly two non-isomorphic rings with p elements.

2.29 Determine (giving a proof or a counter-example) which of the following statements are true.

(*a*) Every subring of a field is an integral domain.

(*b*) Every quotient ring of an integral domain is an integral domain.

(*c*) The invertible elements in an integral domain form a cyclic group under multiplication.

(*d*) If R is a ring with an identity and the non-zero elements of R form a cyclic group under multiplication then R is a field.

(*e*) Given a field F there is an integral domain R with F a subring of R.

2.30 An element x of a ring R is said to be *nilpotent* if $x^n = 0$ for some positive integer n.

(*a*) If R is a commutative ring show that the nilpotent elements of R form an ideal N.

(*b*) Show also that the ring R/N has no non-zero nilpotent elements.

(*c*) If R is the ring $\mathbb{Z}_n$, what is N?

(*d*) If R is not commutative show (by considering for example $R = \mathrm{Mat}_{2\times 2}(\mathbb{R})$) that the nilpotent elements do not necessarily form an ideal.

2.31 Let R be a commutative ring with an identity and let $a_i,\ b_j,\ x \in R$ where

$i = 0, \ldots, n$ and $j = 0, \ldots, m$. Establish the identity

$$\sum_{i=0}^{n} \sum_{j=0}^{m} a_i b_j x^{i+j} = \sum_{k=0}^{n+m} \left(\sum_{j=0}^{k} a'_{k-j} b'_j \right) x^k,$$

where $a'_i = a_i$ for $0 \leqslant i \leqslant n$ and $a'_i = 0$ otherwise, and similarly $b'_j = b_j$ for $0 \leqslant j \leqslant m$ and $b'_j = 0$ otherwise.

Hence deduce that, for every $x \in R$, the *substitution map* $\zeta_x : R[X] \to R$ defined by

$$\zeta_x \left(\sum_{i=0}^{n} a_i X^i \right) = \sum_{i=0}^{n} a_i x^i$$

is a ring morphism.

Given $x \in \mathbb{R}$ let $\zeta_x : \mathbb{R}[X] \to \mathbb{R}$ be the substitution morphism. Determine the image and kernel of ζ_x. Find a polynomial that generates Ker ζ_x.

2.32 Let $\vartheta_m : \mathbb{Z} \to \mathbb{Z}_m$ be the 'natural' morphism, described by

$$\vartheta_m(a) = \text{the remainder in the euclidean division of } a \text{ by } m.$$

Prove that $\bar{\vartheta}_m : \mathbb{Z}[X] \to \mathbb{Z}_m[X]$ described by

$$\bar{\vartheta}_m \left(\sum_{i=0}^{n} a_i X^i \right) = \sum_{i=0}^{n} \vartheta_m(a_i) X^i$$

is a surjective ring morphism. What is the kernel of $\bar{\vartheta}_m$?

Find the quotient and remainder on dividing

(*a*) $X^4 + 3X^3 + 2X^2 + X + 4$ by $3X^2 + 2X$ in the ring $\mathbb{Z}_5[X]$;

(*b*) X^{10} by $X^2 + 1$ in the ring $\mathbb{Z}_2[X]$.

Find all positive integers n such that $X^2 + 2$ divides $X^5 - 10X + 12$ in the ring $\mathbb{Z}_n[X]$.

2.33 (*a*) Find a commutative ring R with an identity for which $R[X]$ has zero divisors. If $f(X) \in R[X]$ has constant term 1, can $f(X)$ be a zero divisor? Can you generalise this result?

(*b*) Let R be a commutative ring with an identity. Show that $1 + aX$ is invertible in $R[X]$ if and only if there exists $n > 0$ with $a^n = 0$.

(*c*) Let F be a field. Find all the invertible elements of $F[X]$.

2.34 If R is an integral domain prove that so also is $R[X]$. If R is a field, prove that $R[X]$ is never a field.

Prove that any polynomial in $\mathbb{Z}_4[X]$ which has a multiplicative inverse has constant term either 1 or 3. If

$$p(X) = a_0 + a_1X + \cdots + a_nX^n$$

satisfies $a_0 = 1$ or 3 and $a_i = 0$ or 2 for $i \geqslant 1$, prove that $p(X)$ is invertible.

2.35 Show by means of counter-examples that the division algorithm fails in (*a*) $\mathbb{Z}_4[X]$ and (*b*) $\mathbb{Z}[X]$.

2.36 Find the highest common factor of each of the following pairs of polynomials and express it in the form $a(X)f(X) + b(X)g(X)$:

(*a*) $f(X) = X^3 - 1$, $g(X) = X^4 - X^3 + X^2 + X - 2$ in the ring $\mathbb{R}[X]$;

(*b*) $f(X) = X^2 + 1, g(X) = X^3 + 2X - \mathrm{i}$ in the ring $\mathbb{C}[X]$;

(*c*) $f(X) = X^4 + 1, g(X) = X^2 + X + 2$ in the ring $\mathbb{Z}_3[X]$.

2.37 Consider the set F of matrices of the form $aI_2 + bK$ where I_2 is the 2×2 identity matrix and

$$K = \begin{bmatrix} 1 & 1 \\ -1 & 0 \end{bmatrix}.$$

Prove that F is a subfield of the ring $\mathrm{Mat}_{2\times 2}(\mathbb{R})$.

Show that the polynomial $X^2 - X + 1$ is irreducible over $\mathbb{Q}$ and prove that the fields $\mathbb{Q}[X]/(X^2 - X + 1)$ and F are isomorphic.

2.38 Show that the set R of real matrices of the form

$$\begin{bmatrix} a & b & c \\ 0 & d & e \\ 0 & 0 & f \end{bmatrix}$$

forms a subring of the ring $\mathrm{Mat}_{3\times 3}(\mathbb{R})$.

Show that the mapping $\alpha : R \to \mathbb{R} \times \mathbb{R} \times \mathbb{R}$ given by

$$\begin{bmatrix} a & b & c \\ 0 & d & e \\ 0 & 0 & f \end{bmatrix} \to (a, d, f)$$

is a ring morphism. Find an ideal I of R with $R/I \cong \mathrm{Im}\, \alpha$.

2.39 Compute the product

$$\begin{bmatrix} 3 & -1 \\ -5 & 1 \\ 1 & 1 \end{bmatrix} \begin{bmatrix} a & b \\ c & d \end{bmatrix} \begin{bmatrix} 2 & 1 & 1 \\ 1 & 1 & 2 \end{bmatrix}.$$

Deduce that $\vartheta : \mathrm{Mat}_{2\times 2}(\mathrm{IR}) \to \mathrm{Mat}_{3\times 3}(\mathrm{IR})$ given by

$$\vartheta \begin{bmatrix} a & b \\ c & d \end{bmatrix} =$$

$$\frac{1}{2}\begin{bmatrix} 6a+3b-2c-d & 3a+3b-c-d & 3a+6b-c-2d \\ -10a-5b+2c+d & -5a-5b+c+d & -5a-10b+c+2d \\ 2a+b+2c+d & a+b+c+d & a+2b+c+2d \end{bmatrix}$$

is a ring morphism. Show that $\mathrm{Im}\,\vartheta \cong \mathrm{Mat}_{2\times 2}(\mathrm{IR})$.

2.40 Find the highest common factor $h(X)$ in the ring $\mathbb{Q}[X]$ of

$$f(X) = X^5 + X^4 + 2X^3 + 2X^2 + 2X + 1 \quad \text{and}$$
$$g(X) = X^4 + X^2 + 1.$$

Show that the smallest ideal I of $\mathbb{Q}[X]$ which contains $f(X)$ and $g(X)$ is $\{k(X)h(X) \mid k(X) \in \mathbb{Q}[X]\}$.

Prove that $\mathbb{Q}[X]/I$ is isomorphic to the subring $S = \{a + b\sqrt{3}\,\mathrm{i} \mid a, b \in \mathbb{Q}\}$ of $\mathbb{C}$.

2.41 Consider the polynomials

$$f(X) = X^4 + X^3 + 4X^2 + 11X - 2,$$
$$g(X) = X^3 + X^2 + 5X - 2.$$

Prove that in $\mathbb{Z}_5[X]$ the highest common factor of $f(X), g(X)$ is $X^2 + 2X + 2$ whereas in $\mathbb{Z}_7[X]$ it is $X + 4$.

2.42 Given $q \in \mathbb{Q}$ let S_q be the set of real matrices of the form $aI_2 + bK$ where $a, b \in \mathbb{Q}$ and

$$K = \begin{bmatrix} 0 & q \\ 1 & 0 \end{bmatrix}.$$

Show that S_q is a subring of $\mathrm{Mat}_{2\times 2}(\mathbb{Q})$ that is commutative and has an identity. Prove that S_q is a field if and only if $\sqrt{q} \notin \mathbb{Q}$. In this case show that $S_q \cong \mathbb{Q}[X]/(X^2 - q)$.

2.43 Let R be a ring with an identity and let I be an ideal of R. If I contains an invertible element of R, prove that $I = R$.

Suppose now that R is the ring of upper triangular real 2×2 matrices, i.e. matrices of the form

$$\begin{bmatrix} x & y \\ 0 & z \end{bmatrix}$$

where $x, y, z \in \mathbb{R}$. Prove that

$$I_1 = \left\{ \begin{bmatrix} x & y \\ 0 & 0 \end{bmatrix} \mid x, y \in \mathbb{R} \right\}$$

$$I_2 = \left\{ \begin{bmatrix} 0 & y \\ 0 & z \end{bmatrix} \mid y, z \in \mathbb{R} \right\}$$

$$I_3 = \left\{ \begin{bmatrix} 0 & y \\ 0 & 0 \end{bmatrix} \mid y \in \mathbb{R} \right\}$$

are ideals of R and that, apart from R and $\{0\}$, these are the only ideals of R.

2.44 Let $f(X) \in \mathbb{Z}_2[X]$ be irreducible with $\deg f(X) \geqslant 2$. Prove that $f(X)$ has an odd number of non-zero coefficients.

Find all the irreducible polynomials in $\mathbb{Z}_2[X]$ of degree less than or equal to 4. Hence give examples of fields with 4, 8 and 16 elements.

Use the above results to write the following polynomials in $\mathbb{Z}_2[X]$ as products of irreducible polynomials:

(*a*) $X^5 + X^4 + 1$;
(*b*) $X^6 + X^5 + X^2 + X + 1$;
(*c*) $X^6 + X^4 + X^3 + X^2 + 1$;
(*d*) $X^7 + X^6 + X^4 + X^3 + 1$;
(*e*) $X^7 + X^6 + X^3 + X + 1$;
(*f*) $X^8 + X^6 + X^5 + X^4 + X^3 + X^2 + 1$.

Show that $\mathbb{Z}_2[X]/(X^4 + X + 1)$ is a field in which the multiplicative group of non-zero elements is cyclic and generated by the image of X under the natural morphism from $\mathbb{Z}_2[X]$ to $\mathbb{Z}_2[X]/(X^4 + X + 1)$.

Show that the image of X in $\mathbb{Z}_2[X]/(X^4 + X^3 + X^2 + X + 1)$ does not generate the multiplicative group of non-zero elements but, by finding a generator, show that this group is cyclic.

Solutions to Chapter 1

1.1 If $[a]=[x]$ and $[b]=[y]$ then $a-x=np$ and $b-y=nq$ for some $p, q \in \mathbb{Z}$. Then

$$ab-xy = ab - xb + xb - xy$$
$$= (a-x)b + x(b-y) = n(pb+qx)$$

and so $[ab]=[xy]$. Similarly

$$a+b-(x+y) = np + nq = n(p+q)$$

gives $[a+b]=[x+y]$.

$[2][x]=[3]$ has no solution.

$[5][x]=[0]$ has precisely one solution, namely $[0]$.

$[2][x]=[2]$ has two solutions, namely $[1]$ and $[4]$.

1.2 Observe that $n^2 \equiv (n+6k)^2 (\text{mod } 6)$ for all integers n and k, so that $nR(n+6k)$. Thus we need inspect only the equivalence classes of the integers 0,1,2,3,4,5. Now it is readily seen that, modulo 6, we have

$$0^2 \equiv 0, \quad 1^2 \equiv 1 \equiv 5^2, \quad 3^2 \equiv 3, \quad 4^2 \equiv 4 \equiv 2^2.$$

There are therefore only four R-classes, namely

$$[0]_R = \{6k \mid k \in \mathbb{Z}\};$$
$$[1]_R = \{6k+1, 6k+5 \mid k \in \mathbb{Z}\};$$
$$[2]_R = \{6k+2, 6k+4 \mid k \in \mathbb{Z}\};$$
$$[3]_R = \{6k+3 \mid k \in \mathbb{Z}\}.$$

That $[x]_R [y]_R = [xy]_R$ defines a binary operation follows from the observation that if aRx and bRy then $a^2 = x^2 + 6p$ and $b^2 = y^2 + 6q$ whence $a^2b^2 = (x^2+6p)(y^2+6q) = x^2y^2 + 6r$ where $r \in \mathbb{Z}$, and so $abRxy$. The corresponding Cayley table is then

$\cdot$	$[0]_R$	$[1]_R$	$[2]_R$	$[3]_R$
$[0]_R$	$[0]_R$	$[0]_R$	$[0]_R$	$[0]_R$
$[1]_R$	$[0]_R$	$[1]_R$	$[2]_R$	$[3]_R$
$[2]_R$	$[0]_R$	$[2]_R$	$[2]_R$	$[0]_R$
$[3]_R$	$[0]_R$	$[3]_R$	$[0]_R$	$[3]_R$

That $[x]_R + [y]_R = [x+y]_R$ does not define a binary operation on S follows from the observation that if aRx and bRy then $a^2 = x^2 + 6p$ and $b^2 = y^2 + 6q$ and so

$$(a+b)^2 = (x+y)^2 - 2xy + 2ab + 6(p+q)$$

from which we see that $(a+b)R(x+y)$ if and only if $ab \equiv xy \pmod 3$. This is not always the case; take, for example $a = b = x = 1, y = -1$.

1.3 The Cayley table for $*$ is readily computed and is the following:

$*$	2	3	5	7	11	13
2	2	3	5	7	11	13
3	3	2	3	2	3	7
5	5	3	2	5	7	2
7	7	2	5	3	2	3
11	11	3	7	2	5	11
13	13	7	2	3	11	3

It is clear from this table that 2 is an identity element. Moreover, every element has an inverse:

2 has inverse 2;
3 has inverses 3 and 7;
5 has inverses 5 and 13;
7 has inverses 3 and 11;
11 has inverse 7;
13 has inverse 5.

Finally, $*$ is not associative since with respect to an associative law no element can have more than one inverse.

1.4 That the operation $*$ is associative follows from a purely routine verification of the associative law. In searching for an identity element we look for an element (x', y', z') such that $(x, y, z) * (x', y', z') = (x, y, z)$ for all x, y, z;

i.e. we have to solve

$$xx' = x; \quad \alpha y' + yx' = y; \quad \alpha z' + zx' = z.$$

Clearly, $x' = 1$ and $y' = z' = 0$ satisfy these requirements. Consequently $(1, 0, 0)$ is the identity element with respect to $*$.

Now (x, y, z) has an inverse (x', y', z') if and only if $(x, y, z) * (x', y', z') = (1, 0, 0)$; so given (x, y, z) we have to solve the equations

$$xx' = 1; \quad \alpha y' + yx' = 0; \quad \alpha z' + zx' = 0.$$

These equations can be written in matrix form

$$\begin{bmatrix} x & 0 & 0 \\ y & \alpha & 0 \\ z & 0 & \alpha \end{bmatrix} \begin{bmatrix} x' \\ y' \\ z' \end{bmatrix} = \begin{bmatrix} 1 \\ 0 \\ 0 \end{bmatrix}$$

and we know from matrix theory that a unique solution exists if and only if the coefficient matrix A is such that $\det A \neq 0$. Now clearly $\det A = x\alpha^2$. Hence a solution exists if and only if x and α are non-zero, in which case the solution is given by

$$x' = \frac{1}{x}, \quad y' = -\frac{y}{\alpha x}, \quad z' = -\frac{z}{\alpha x}.$$

1.5

(*a*) Associative; identity is 0; inverse of $2n$ is $-2n$; a group.
(*b*) $(1-1)-1 \neq 1-(1-1)$ so not associative; no identity; not a group.
(*c*) Associative; identity is 1; 0 has no inverse; not a group.
(*d*) Associative; identity is 1; inverse of $\frac{r}{s}$ is $\frac{s}{r}$; a group.
(*e*) Associative operation on $\mathbb{Z}_6$ but $\mathbb{Z}_6 \setminus \{[0]\}$ is not closed under the operation; for example, $[2][3] = [0]$.
(*f*) Associative; identity is $[1]$; $[0]$ has no inverse; not a group.
(*g*) Associative; identity is 1; inverse of $e^{2n\pi i/3}$ is $e^{-2n\pi i/3}$; a group.

1.6

(*a*) No : 1 is the identity but $2 * a = 1$ implies $\max\{2, a\} = 1$ which is impossible.
(*b*) No : there is no identity element (if there were such an element e then from $a * e = a$ we would have $\min\{a, e\} = a$ whence $a \leqslant e$ for all $a \in \mathbb{Z}$ and e would be the greatest element of $\mathbb{Z}$).
(*c*) No : 0 is the identity but 1 has no inverse since $1 * b = 0$ implies $1 + b - b = 0$ which is nonsense.

(*d*) No : the operation is not associative; for example, we have $2 * (3 * 4) = 2 * 1 = 1$ and $(2 * 3) * 4 = 1 * 4 = 3$.

(*e*) Yes : $(0, 0)$ is the identity; $(-a, -b)$ is the inverse of (a, b).

(*f*) No: the operation is not associative; for example we have $((1, 2) * (2, 3)) * (2, 1) = (25, 17)$ and $(1, 2) * ((2, 3) * (2, 1)) = (17, 15)$.

(*g*) Yes : $(1, 0)$ is the identity and the inverse of (a, b) is

$$\left(\frac{a}{a^2 + b^2}, -\frac{b}{a^2 + b^2}\right).$$

(Think about the non-zero complex numbers under multiplication!)

1.7 It is clear that + is a law of composition on T which is associative since addition of matrices is associative. The identity element of T is the 2×2 zero matrix and the additive inverse of $\begin{bmatrix} x & y \\ -y & x \end{bmatrix}$ is $\begin{bmatrix} -x & -y \\ y & -x \end{bmatrix}$. Thus T is a group under +. Now

$$\begin{bmatrix} x & y \\ -y & x \end{bmatrix}\begin{bmatrix} x' & y' \\ -y' & x' \end{bmatrix} = \begin{bmatrix} xx' - yy' & xy' + yx' \\ -(xy' + yx') & xx' - yy' \end{bmatrix} \in T$$

so multiplication is a law of composition on T which is associative since multiplication of matrices is associative. Clearly, the 2×2 identity matrix is the identity element of T. Now $\det \begin{bmatrix} x & y \\ -y & x \end{bmatrix} = x^2 + y^2$ and this is 0 if and only if $x = y = 0$. Thus the set of invertible elements of T under multiplication is $T \setminus \{0\}$, which is then a group.

1.8 Suppose that $f : x \to ax + b (a \neq 0)$ and that $g : x \to a^*x + b^* (a^* \neq 0)$. Then

$$f[g(x)] = f(a^*x + b^*) = a(a^*x + b^*) + b = aa^*x + ab^* + b$$

where $aa^* \neq 0$. Thus $f \circ g \in G$ and so the mapping $(f, g) \to f \circ g$ describes a law of composition on G which is clearly associative. Also it is evident that G has an identity element under this law, namely the identity map $\mathrm{id}_{\mathbb{R}}$ on $\mathbb{R}$.

With f and g as above, we have that $f \circ g = \mathrm{id}_{\mathbb{R}} = g \circ f$ if and only if, for all $x \in \mathbb{R}$,

$$a^*(ax + b) + b^* = x = a(a^*x + b^*) + b,$$

i.e. if and only if $aa^* = 1$ and $a^*b + b^* = 0 = ab^* + b$. This is the case if and only if

$$a^* = \frac{1}{a} \quad \text{and} \quad b^* = -\frac{b}{a}.$$

Since $a \neq 0$ implies $\frac{1}{a} \neq 0$, we have that

$$g : x \to \frac{1}{a}x - \frac{b}{a}$$

is the inverse of f in G. Hence G is a group.

1.9 The operation $*$ is associative. To see this, we must check several cases. For example, suppose that $a < 0$ and $b < 0$; then we have

$$(a * b) * c = \frac{a}{b} * c = \frac{ac}{b} \quad \text{and} \quad a * (b * c) = a * \frac{b}{c} = \frac{ac}{b}.$$

Similar arguments in the cases $a < 0$, $b > 0$ and $a > 0$, $b < 0$ and $a > 0$, $b > 0$ complete the proof of associativity.

1 is the identity element. In fact $1 * a = 1a = a$; and for $a > 0$ we have $a * 1 = a1 = a$, while for $a < 0$ we have

$$a * 1 = \frac{a}{1} = a.$$

If $a > 0$ then $\frac{1}{a}$ is its inverse; for we have

$$a * \frac{1}{a} = a\frac{1}{a} = 1, \quad \frac{1}{a} * a = \frac{1}{a}a = 1.$$

If $a < 0$ then a is its own inverse; for we have

$$a * a = \frac{a}{a} = 1.$$

Hence the operation $*$ makes $\mathbb{R} \setminus \{0\}$ into a group.

1.10 See Fig. S1.1.

(*a*) The only symmetries are the identity and the reflection in the broken line shown. Hence the group of symmetries is the same as the symmetric group S_2.

(*b*) The group of symmetries is the same as that of the non-square rectangle, as is shown by the diagram. Thus it is the Klein 4-group V_4.

Solutions to Chapter 1

Fig.S1.1

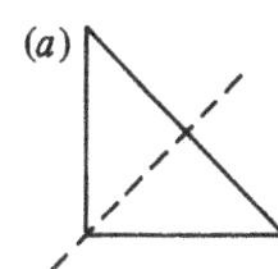

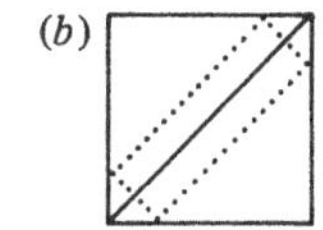

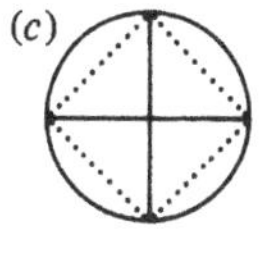

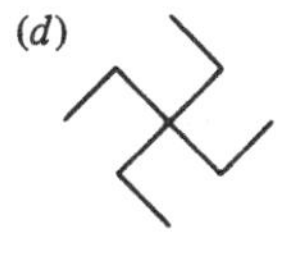

(*c*) The group of symmetries is the same as that of the square, as is shown by the diagram. Thus it is the dihedral group D_4.

(*d*) The only symmetries of the swastika are the rotations through multiples of $\pi/2$. Thus it is clear that the group of symmetries is C_4.

1.11 Given an n-sided regular polygon, label a pair of adjacent vertices A and B. In performing a symmetry there are n vertices to which we can move A; and once the position of A has been fixed there are two (adjacent) vertices from which to choose the position of B. This then determines the position of the entire polygon. Thus we see that in all there are $2n$ symmetries.

These symmetries consist of

(*a*) rotations through $(2\pi/n)k$ where $k = 0, 1, \ldots, n-1$;

(*b*) reflections in lines joining the centre of the polygon to its vertices, or to the mid-points of its edges. (These are the same when n is odd.)

Since (*a*) and (*b*) account for n symmetries each, they describe all of them.

To see that the group is non-abelian let A and B be adjacent vertices. Consider the rigid motions f, g described respectively by a reflection in the line joining A to the centre of the polygon, and a rotation through $2\pi/n$. Then $f \circ g$ and $g \circ f$ have different actions on B. Draw pictures to see this!

1.12 Consider an adjacent pair of symbols. If a symmetry leaves this pair in the same order then clearly it is of the form a^n where n is the number of units in the translation (negative values of n represent translations to the left). On the other hand, if a symmetry reverses the order of the pair then clearly it is of the form ba^n. Now to be able to multiply two symmetries of the form $b^m a^n$ and obtain another symmetry of this form, it suffices that ab should be of this form; and it is easy to see that indeed $ab = ba^{-1}$. Hence the group of symmetries is generated by a and b subject to the relations $b^2 = 1$, $ab = ba^{-1}$.

There are several ways of exhibiting the seven different possible patterns (either with symbols, letters, pictures, etc.). Here is one way:

(1) The symmetries of the pattern

$\cdots \subseteq\subseteq\subseteq\subseteq\subseteq \cdots$

consist of translations only.

(2) The symmetries of the pattern

$\cdots \int\int\int\int\int \cdots$

consist of translations and rotations.

(3) The symmetries of the pattern

· · · ƎƎƎƎƎ · · ·

consist of translations and horizontal reflections.

(4) The symmetries of the pattern

· · · VVVVV · · ·

consist of translations and vertical reflections.

(5) The symmetries of the pattern

· · · XXXXX · · ·

consist of translations, rotations, and both horizontal and vertical reflections.

(6) The symmetries of the pattern

· · · U∩U∩U · · ·

consist of translations, vertical reflections, and 'slide reflections' (i.e. a translation followed by a horizontal reflection).

(7) The symmetries of the pattern

· · · Ǝ Ǝ Ǝ Ǝ Ǝ · · ·

consist of translations and 'slide reflections'.

1.13 A rotation is fixed when we describe what happens to two adjacent corners. We can put the first corner into one of eight different positions and then rotate the other into one of three adjacent positions. Thus we see that $|G| = 8 \times 3 = 24$.

Consider now rotations x that fix a particular corner. They all satisfy $x^3 = 1$. There are eight corners, but a rotation that fixes one corner also fixes the one diagonally opposite. Thus we can find eight different non-identity elements of this kind (two for each pair of opposite corners).

Similarly, rotations of 90° about the centre of a face satisfy $x^4 = 1$. So do rotations of 180° and 270°, and it should be noted that the rotations of 180° satisfy $x^2 = 1$ also. Such rotations send opposite pairs of faces into themselves. So, when we count the non-identity ones, we find that there are nine (namely, three for each pair of opposite faces).

Similarly, rotations of 180° about the mid-point of an edge satisfy $x^2 = 1$, and there are six of these (namely, one for each pair of opposite edges).

Collecting these together, and remembering the identity, we have accounted for $1 + 8 + 9 + 6 = 24$ elements of G, i.e. all of G.

It is an easy matter to do the same for the octohedron. Simply note that the centres of the faces of a cube lie at the corners of an octohedron (draw a picture!), so a rotation of the cube gives a unique rotation of its inscribed octohedron, and vice versa.

(*Remark.* The octohedron is called the 'dual' of the cube. The dual of the icosohedron is the dodecahedron. You may wish to describe in a similar way the elements of its rotation group. In this case we have $|G| = 60$.)

1.14 We use the fact (which we shall refer to as $(*)$) that every element of the group appears once and once only in each row and column in the Cayley table. It is immediate from this that $ay = z$ and $az = x$, which completes the second row. As for the third row, we observe that $a^2 = b$ and so $a^3 = ab = ba$. But it is given that $ab = e$; hence $ba = e$. Also, $bb = a^2a^2 = a^4 = a^3a = ea = a$. We can now use $(*)$ to complete the third row as well as the second and third columns. Consider now the element xx. We have clearly $xx \in \{a, b, e\}$. Since $xz = a$ we cannot have $xx = a$; and if $xx = b$ we have the contradiction $xb = x^3 = bx$. Hence we must have $xx = e$. It follows that $xy = xxb = eb = b$ and $yx = axx = ae = a$. The rest follow by $(*)$. The complete table is

	e	a	b	x	y	z
e	e	a	b	x	y	z
a	a	b	e	y	z	x
b	b	e	a	z	x	y
x	x	z	y	e	b	a
y	y	x	z	a	e	b
z	z	y	x	b	a	e

1.15 If k is the order of $g \in G$ then g generates a cyclic subgroup of order k and hence, by Lagrange's theorem, $k|n$. Also, $g^k = 1$ and hence $g^n = 1$.

If an integer $a \equiv 0 (\text{mod } p)$ then it is self-evident that $a^p \equiv a (\text{mod } p)$. If $a \not\equiv 0 (\text{mod } p)$, calculate in the group $\mathbb{Z}_p \setminus \{0\}$. This is a group of order $p - 1$ and hence $a^{p-1} = 1$ by the above. Consequently, $a^p = a$.

(*a*) Since 11 is prime, we have $7^{10} \equiv 1 (\text{mod } 11)$ and so $7^{100} \equiv 1^{10} \equiv 1$ (mod 11).

(*b*) Since 13 is prime, we have $9^{12} \equiv 1 \pmod{13}$ and so $9^{36} \equiv 1^3 \equiv 1 \pmod{13}$, whence $9^{37} \equiv 9 \pmod{13}$.

1.16 (*a*) From the information given, the result is true for $n = 1$. This then anchors the induction. As for the inductive step, we have (supposing that the result holds for n)

$$a^{n+1}b = aa^n b = aba^{kn} = ba^k a^{kn} = ba^{k(n+1)}.$$

(*b*) Again, from the information given, the result is true for $n = 1$. The inductive step in this case is as follows:

$$\begin{aligned} ab^{n+1} = ab^n b &= b^n a^{k^n} b \\ &= b^n b a^{kk^n} \text{ by } (a) \\ &= b^{n+1} a^{k^{n+1}}. \end{aligned}$$

That $(ba^n)^2$ commutes with a can be seen, for example, from the equations

$$a(ba^n)^2 = aba^n ba^n = abba^{nk}a^n = a^{nk+n+1}$$
$$(ba^n)^2 a = ba^n ba^n a = bba^{kn}a^n a = a^{kn+n+1}$$

and that it also commutes with b can be seen from

$$b(ba^n)^2 = bba^n ba^n = a^n ba^n = ba^{kn}a^n = ba^{(k+1)n}$$
$$(ba^n)^2 b = ba^n ba^n b = ba^n bba^{kn} = ba^{(k+1)n}.$$

1.17 $a = 1$ or $b = 1$ or $a = b$ contradicts $ab \neq ba$; so $1, a, b$, are distinct. Consider now ab; we cannot have $ab = 1$ (since then $b = a^{-1}$ which contradicts $ab \neq ba$), nor can we have $ab = a$ (for then, by cancellation, $b = 1$). Similarly, $ab \neq b$. Thus $1, a, b, ab, ba$ are distinct.

Now $a^2 \neq a$ (otherwise $a = 1$ by cancellation); $a^2 \neq b$ (otherwise $ab = ba$); $a^2 \neq ab$ (otherwise $a = b$ by cancellation); and $a^2 \neq ba$ (otherwise $a = b$ by cancellation). Thus we see that either $a^2 = 1$ or $a^2 \notin \{1, a, b, ab, ba\}$. Note that the latter implies that G contains at least six elements. Suppose then that the former holds and consider aba. We have that $aba \neq 1$ (otherwise $ab = a^{-1} = ba$); $aba \neq a$ (otherwise $ab = 1$); $aba \neq b$ (otherwise, since $a^2 = 1$ by hypothesis, $ba = a^2 ba = ab$); $aba \neq ab$ (otherwise $a = 1$ by cancellation); and $aba \neq ba$ (otherwise $a = 1$ by cancellation). Thus in this case G also contains at least six elements.

The Cayley table for F under $\circ$ is precisely that given in question 1.14, with a, b, x, y, z replaced by p, q, a, c, b respectively. Thus F is a group under $\circ$.

1.18 (*a*) Since $ab = ba$ we have $b^{-1}ab = b^{-1}ba = a$. Multiplying on the right by b^{-1} we obtain $b^{-1}a = ab^{-1}$. Also, since a and b commute, we have

$$xax^{-1}.xbx^{-1} = xabx^{-1}$$
$$= xbax^{-1} = xbx^{-1}.xax^{-1}.$$

(*b*) If G is abelian than $(ab)^{-1} = b^{-1}a^{-1} = a^{-1}b^{-1}$. Conversely, if this holds for all $a, b \in G$ then writing a^{-1} for a and b^{-1} for b we obtain $ba = ab$.

(*c*) If $(ab)^2 = a^2b^2$ then we have $abab = aabb$ and so, multiplying on the left by a^{-1} and on the right by b^{-1} we obtain $ba = ab$.

(*d*) If $a^2 = 1$ for all $a \in G$ then given any $x, y \in G$ we have $(xy)^2 = 1$, $x^2 = 1$ and $y^2 = 1$. Consequently $(xy)^2 = x^2y^2$ and the result follows by (*c*).

(*e*) The result is trivial for $n = 1$. For the inductive step, suppose that $(x^{-1}yx)^n = x^{-1}y^nx$; then

$$(x^{-1}yx)^{n+1} = (x^{-1}yx)^nx^{-1}yx = x^{-1}y^nxx^{-1}yx = x^{-1}y^{n+1}x.$$

(*f*) We are given that $b^6 = 1$ and $ab = b^4a$. From the second of these and (*e*) we obtain $b = a^{-1}b^4a = (a^{-1}ba)^4$. Thus

$$b^3 = (a^{-1}ba)^{12} = a^{-1}b^{12}a = a^{-1}1a = 1,$$

and $ab = b^4a = b^3ba = ba$.

1.19 The orders are respectively 6,3,4. The element $\begin{bmatrix} 0 & 1 \\ 1 & 0 \end{bmatrix}$ has order 2, as does $\begin{bmatrix} -1 & 0 \\ 0 & -1 \end{bmatrix}$ (and many more!).

The given formula is clearly true for $n = 1$. As for the inductive step, suppose that $A^k = \begin{bmatrix} 1 & k \\ 0 & 1 \end{bmatrix}$; then

$$A^{k+1} = A^kA = \begin{bmatrix} 1 & k \\ 0 & 1 \end{bmatrix}\begin{bmatrix} 1 & 1 \\ 0 & 1 \end{bmatrix} = \begin{bmatrix} 1 & k+1 \\ 0 & 1 \end{bmatrix}.$$

It is clear from this that A does not have finite order. Since the subgroup generated by A is thus infinite, so also is G. Finally, taking $a = A$, $b = I_2$ and $c = \begin{bmatrix} 1 & 0 \\ 1 & 1 \end{bmatrix}$, we see that a and b commute, as do b and c, but a and c do not commute.

1.20 (*a*) Cyclic : every element is a power of $\begin{bmatrix} 1 & 1 \\ 0 & 1 \end{bmatrix}$. The group is then abelian. It is also infinite.

(*b*) Cyclic : every element is a power of [2]. The group is then abelian. It is finite.

(*c*) Not abelian : for example, $(123) \circ (12) \neq (12) \circ (123)$. Thus the group cannot be cyclic. It is finite.

(*d*) Cyclic : every element is a power of (12). The group is then abelian. It is finite.

(*e*) Cyclic : every element is a multiple of 2. The group is then abelian. It is infinite.

(*f*) Abelian but not cyclic : for if it were cyclic then every real number would be a multiple of some $r \in \mathbb{R}$, whence 1 would be a multiple of r and r would be rational; but then, for example, π cannot be a multiple of r. Clearly infinite.

1.21 If $a \neq 0$ and $c \neq 0$ then $ac \neq 0$, so the given prescription defines a law of composition on G. To see that it is associative, check that

$$[(a,b)(c,d)](f,g) = (acf, bcf + df + g) = (a,b)[(c,d)(f,g)].$$

The identity element is $(1, 0)$; and it is readily seen that

$$(a,b)^{-1} = \left(\frac{1}{a}, -\frac{b}{a}\right).$$

Clearly, the group G is non-abelian.

It is easy to check that H, K, M are subgroups of G. For example, if $(y, 0) \in K$ then $(y, 0)^{-1} = (y^{-1}, 0)$ and hence, given $(x, 0)$ and $(y, 0)$ in K, we have

$$(x,0)(y,0)^{-1} = (x,0)(y^{-1},0) = (xy^{-1},0) \in K.$$

As for L, except for the case $n = 0$ (when $L = K$), the subset L does not contain the identity element $(1, 0)$ and so cannot be a subgroup.

$(-1, a)$ is an element of order 2 for every $a \in \mathbb{R}$; for

$$(-1,a)(-1,a) = (1, -a + a) = (1, 0).$$

Suppose now that (a, b) has order 3. Then

$$(1,0) = (a,b)^3 = (a^3, b(a^2 + a + 1))$$

so $a = 1$ and $b = 0$. Thus $(a, b) = (1, 0)$, which has order 1. We conclude that G has no elements of order 3.

1.22 (*a*) If $x, y \in H$ then $x = 2^n$, $y = 2^m$ for some $n, m \in \mathbb{Z}$ and so $xy^{-1} = 2^{n-m} \in H$. Hence H is a subgroup.

If $x, y \in K$ then for $n, m, r, s \in \mathbb{Z}$ we have

$$x = \frac{1+2n}{1+2m}, \quad y = \frac{1+2r}{1+2s}.$$

That K is a subgroup now follows from the fact that

$$xy^{-1} = \frac{1+2n}{1+2m} \cdot \frac{1+2s}{1+2r} = \frac{1+2(n+s+2ns)}{1+2(m+r+2mr)} \in K.$$

(*b*) Check that $[a], [b] \in H$ implies $[a] - [b] \in H$.

(*c*) The subgroups are

$$\{e\}, \quad \{e, x\}, \quad \{e, y\}, \quad \{e, z\},$$
$$\{e, a, b\}, \quad \{e, a, b, x, y, z\}.$$

(Note that all subgroups except the whole group are cyclic.)

1.23 It is a routine matter to show that

$$(x_1, y_1)[(x_2, y_2)(x_3, y_3)] = (x_1x_2x_3, (x_1+y_1)(x_2+y_2)(x_3+y_3) - x_1x_2x_3)$$

whence associativity follows by symmetry. The identity element is readily seen to be $(1, 0)$. Also, every element has an inverse, that of (x, y) being

$$(x, y)^{-1} = \left(\frac{1}{x}, -\frac{y}{x(x+y)}\right).$$

Thus S is a group.

To verify that P is a subgroup of S, let $x = (1, a) \in P$ and $y = (1, b) \in P$. Then

$$xy^{-1} = (1, a)\left(1, -\frac{b}{1+b}\right) = \left(1, \frac{1+a}{1+b} - 1\right)$$

and this belongs to P since $1 + a > 0$ and $1 + b > 0$.

1.24 G is, by definition, the set of all 2×2 matrices whose elements are integers and whose determinant is 1. It is an easy matter to show that G is a group under multiplication.

The inverse of $\begin{bmatrix} a & b \\ c & d \end{bmatrix} \in G$ is $\begin{bmatrix} d & -b \\ -c & a \end{bmatrix}$. To show that H_N is a subgroup, let $\begin{bmatrix} a & b \\ c & d \end{bmatrix}$ and $\begin{bmatrix} a' & b' \\ c' & d' \end{bmatrix}$ be in H_N; then

$$\begin{bmatrix} a & b \\ c & d \end{bmatrix}\begin{bmatrix} d' & -b' \\ -c' & a' \end{bmatrix} = \begin{bmatrix} ad' - bc' & -ab' + ba' \\ cd' - dc' & -cb' + da' \end{bmatrix}$$

Fig.S1.2

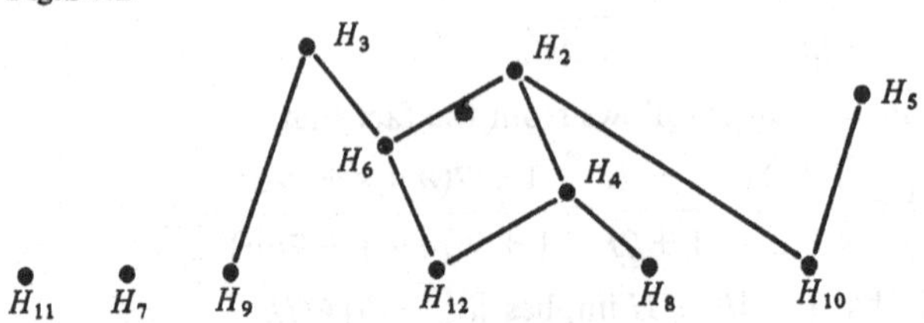

and, using the result of question 1.1, it is easily seen from the definition of H_N that this product matrix belongs to H_N.

The Hasse diagram for the set $\{H_N \mid 2 \leqslant N \leqslant 12\}$ is as shown in Fig. S1.2.

1.25 Let G denote the multiplicative group $\mathbb{Z}_p \setminus \{0\}$.

(*a*) If $a, b \in Q$ then there exist $x, y \in G$ such that $x^2 = a$ and $y^2 = b$. Since G is abelian, we have $(xy^{-1})^2 = x^2(y^2)^{-1} = ab^{-1}$, whence $ab^{-1} \in Q$.

(*b*) Consider a few examples to see what is happening (for example, $p = 3, 5, 7$). You will soon observe that the only time two distinct numbers have the same square is when one is minus the other (for, in $\mathbb{Z}_p$ we have $x^2 = y^2$ if and only if $(x - y)(x + y) = 0$). It follows that only half the numbers in G give distinct squares, so we can take for Q the set $\{1, 2, \ldots, \frac{1}{2}(p-1)\}$. Hence $|Q| = \frac{1}{2}(p-1)$.

(*c*) If $nq \in Q$ then $nq = x^2$ for some x. Since $q = y^2$ for some y, we obtain the contradiction $n = (xy^{-1})^2$.

(*d*) Note that for every $b \in G$ we can solve the equation $mx = b$. Now $mq \notin Q$ for $q \in Q$ (by (*c*)). If then we had $mn \notin Q$ for some $n \notin Q$ then there would not be enough elements of the form mx to 'go round' Q. Hence we must have $mn \in Q$.

(*e*) By the first part of question 1.15 (which is often referred to as the corollary to Lagrange's theorem) and (*b*) above, we have $a^{(p-1)/2} \equiv 1 \pmod{p}$.

1.26 To check that M is a subgroup of the group of all 2×2 real invertible matrices, note that $M(0) = I$ and that $M(\vartheta)M(\varphi) = M(\vartheta + \varphi)$. Hence, given $M(\vartheta)$ and $M(\varphi)$ we have

$$M(\vartheta)[M(\varphi)]^{-1} = M(\vartheta)M(-\varphi) = M(\vartheta - \varphi) \in M.$$

Also, $M(\vartheta)M(\varphi) = M(\vartheta + \varphi)$ which clearly implies that M is abelian.

Observe now that $M\left(\frac{2\pi}{n}\right)$ is an element of order n. In fact

$$M\left(n\left(\frac{2\pi}{n}\right)\right) = M(2\pi) = M(0) = I$$

and if $0 < m < n$ then

$$\left[M\left(\frac{2\pi}{n}\right)\right]^m = M\left(\frac{2\pi m}{n}\right)$$

which is not the identity. Thus $\left\langle M\left(\frac{2\pi}{n}\right)\right\rangle$ is a cyclic group of order n. Hence M has cyclic subgroups of every order n.

The answer to the last part is 'yes' : $M(1)$ generates an infinite cyclic subgroup. To see this, observe that if $[M(\alpha)]^n = I$ then $n\alpha = 2\pi m$ for some integer m so that α is a rational multiple of 2π. Since clearly 1 is not a rational multiple of 2π it follows that $M(1)$ is of infinite order.

1.27 If $G = \langle g \rangle$ is of order n then the subgroups of G are of the form $\langle g^t \rangle$.

Suppose that hcf $\{r, n\} = d \neq 1$. Since $g^r = g^{d(r/d)}$ we have that $g^r \in \langle g^d \rangle$. Since $d = n/t$ for some $t \neq n$ the subgroup $\langle g^d \rangle$ is of order $t \neq n$. Thus $g^r \in \langle g^d \rangle \neq G$ so $\langle g^r \rangle$ is not the whole of G. We conclude from this argument that the other generators of G are g^r where hcf $\{r, n\} = 1$.

By the above, the generators of $\mathbb{Z}_{18}$ are 1,5,7,11,13,17. Consider the subgroups generated by the remaining elements.

$$\langle 2 \rangle = \{0, 2, 4, 6, 8, 10, 12, 14, 16\},$$

a subgroup of order 9. It has generators of the form $r2$ where r, 9 are coprime. Thus $r = 1, 2, 4, 5, 7, 8$ and so $r2 = 2, 4, 8, 10, 14, 16$.

$$\langle 6 \rangle = \{0, 6, 12\},$$

of order 3. Its generators are of the form $r6$ where r, 3 are coprime. So $r = 1, 2$ and $r6 = 6, 12$.

Fig.S1.3

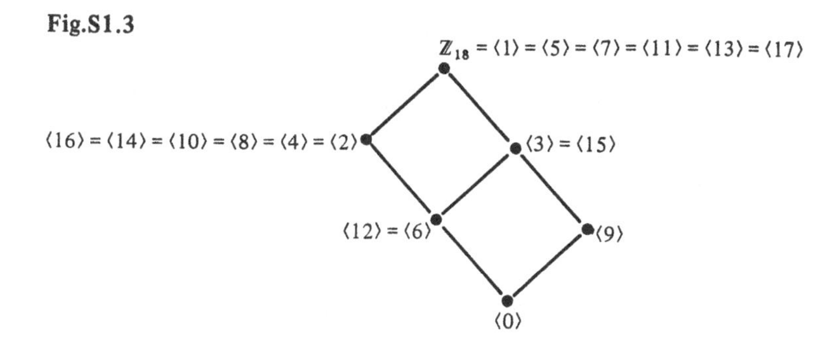

The only subgroups not considered so far are those generated by 3,9,15. Now

$$\langle 3 \rangle = \{0, 3, 6, 9, 12, 15\},$$

of order 6. The generators are $r3$ where r, 6 are coprime. So $r = 1, 5$ and $r3 = 3, 15$. Finally

$$\langle 9 \rangle = \{0, 9\}$$

and the subgroup Hasse diagram is as shown in Fig. S1.3.

1.28 Suppose that $x, y \in H \cap K$. Then $x, y \in H$ gives $xy^{-1} \in H$ and similarly $x, y \in K$ gives $xy^{-1} \in K$. Thus $xy^{-1} \in H \cap K$ and $H \cap K$ is a subgroup.

Let $g \in H \cap K$. Then the order of g divides both $|H|$ and $|K|$. Since by hypothesis we have hcf $\{|H|, |K|\} = 1$ it follows that the order of g divides 1, whence $g = 1$. Thus $H \cap K = \{1\}$.

Clearly $m\mathbb{Z} \cap n\mathbb{Z} = k\mathbb{Z}$ where $k = \text{lcm}\,\{m, n\}$.

1.29 Let $x, y \in C(a)$. Then, using the result of question 1.18(*a*), we have

$$xy^{-1}.a = x.y^{-1}a = x.ay^{-1} = xa.y^{-1} = ax.y^{-1} = a.xy^{-1}$$

and so $xy^{-1} \in C(a)$. Hence $C(a)$ is a subgroup of G.

$C(H) = \bigcap_{a \in H} C(a)$ and so $C(H)$ is a subgroup of G since the intersection of any family of subgroups is a subgroup.

(*a*) $G = H = C_2$.

(*b*) $G = H = S_3$.

(*c*) $G = C_4$ and $H = \langle a^2 \rangle$ where C_4 consists of powers of a.

1.30 Note first that $ab \neq 1$; for if $ab = 1$ then we have the contradiction $a = b^{-1} = b$. Now since G is abelian we have $(ab)^2 = a^2b^2 = 1.1 = 1$.

Let $H = \{1, a, b, ab\}$. To show that H is a subgroup of G it is enough to show that it is stable under multiplication. This is easily seen to be the case since $a^2 = 1 = b^2$ and G is abelian. Now 1 has order 1 and a, b, ab all have order 2. Since $|H| = 4$ it is clear that H cannot be cyclic.

$[2^k - 1]$ has order 2; for, $k > 1$ so

$$(2^k - 1)^2 = 2^{2k} - 2^{k+1} + 1 \equiv 1 (\text{mod } 2^{k+1}).$$

Similarly $(2^k + 1)^2 \equiv 1 (\text{mod } 2^{k+1})$ shows that $[2^k + 1]$ is of order 2. But $[2^k - 1] \neq [2^k + 1]$ for $k > 1$ and so, since G is abelian, we have that

$$H = \{[1], [2^k - 1], [2^k + 1], [2^k - 1][2^k + 1]\}$$

is not cyclic. But every subgroup of a cyclic group is cyclic. So if G were cyclic we would have the contradiction that H is cyclic. We conclude that G cannot be cyclic.

1.31 We observe first that

$$r_1 = \frac{p_1}{q_1} = p_1 q_2 \cdot \frac{1}{q_1 q_2} \in \left\langle \frac{1}{q_1 q_2} \right\rangle$$

and similarly

$$r_2 \in \left\langle \frac{1}{q_1 q_2} \right\rangle, \text{ so that } \langle r_1, r_2 \rangle \subseteq \left\langle \frac{1}{q_1 q_2} \right\rangle$$

Suppose now that $r_i = \dfrac{p_i}{q_i} \in \mathbb{Q}$ for $i = 1, \ldots, n$. Then we have

$$r_i = p_i q_1 q_2 \ldots q_{i-1} q_{i+1} \ldots q_n \cdot \frac{1}{q_1 q_2 \ldots q_n} \in \left\langle \frac{1}{q_1 q_2 \ldots q_n} \right\rangle$$

and so we have that

$$\langle r_1, r_2, \ldots, r_n \rangle \subseteq \left\langle \frac{1}{q_1 q_2 \ldots q_n} \right\rangle.$$

Now the group on the right is cyclic (being generated by a single element); and every subgroup of a cyclic group is cyclic. Hence $\langle r_1, r_2, \ldots, r_n \rangle$ is cyclic.

Clearly, each H_i is cyclic. We have to show that $H_i \subseteq H_{i+1}$ with $H_i \neq H_{i+1}$. Now

$$\frac{1}{i!} = (i+1) \cdot \frac{1}{(i+1)!} \in H_{i+1}.$$

Hence $H_i \subseteq H_{i+1}$; and the inclusion is strict since

$$\frac{1}{(i+1)!} \notin H_i.$$

To show that $\mathbb{Q} = \cup_{n \geqslant 1} H_n$ we must show that, given any rational $\dfrac{a}{b}$ with $b > 0$, we have

$$\frac{a}{b} \in H_i \text{ for some } i.$$

But this follows from the observation that

$$\frac{a}{b} = (b-1)!a \cdot \frac{1}{b!} \in H_b.$$

Finally, suppose that $\mathbb{Q}$ were cyclic, say $\mathbb{Q} = \left\langle \dfrac{a}{b} \right\rangle$. Then

$$\mathbb{Q} = \left\{ \frac{na}{b} \mid n \in \mathbb{Z} \right\} \subseteq H_b$$

which gives $\mathbb{Q} = H_b$, contradicting the fact that the chain of subgroups H_i is infinite.

1.32 Show that $(a) \Rightarrow (e) \Rightarrow (d) \Rightarrow (b) \Rightarrow (c) \Rightarrow (f) \Rightarrow (a)$.

$(a) \Rightarrow (e)$ If $xH = Hx$ then for every $h \in H$ we have $xh \in xH = Hx$ so that there exists $h' \in H$ with $xh = h'x$.

$(e) \Rightarrow (d)$ $x^{-1}h = h'x^{-1}$ gives $x^{-1}hx = h' \in H$.

$(d) \Rightarrow (b)$ This is obvious.

$(b) \Rightarrow (c)$ Since $x^{-1}Hx \subseteq H$ for every $x \in G$ we have (writing x^{-1} for x) that $(x^{-1})^{-1}Hx^{-1} \subseteq H$, i.e. that $xHx^{-1} \subseteq H$. The reverse inclusion comes from the fact that $xhx^{-1} = h' \in H$ gives $h = x^{-1}h'x \in x^{-1}Hx$.

$(c) \Rightarrow (f)$ If $h \in H$ then $x^{-1}hx = h'$ gives $h^{-1}x^{-1}hx = h^{-1}h' \in H$.

$(f) \Rightarrow (a)$ If $h^{-1}x^{-1}hx = h' \in H$ then we have $hx = xhh'$. Thus if $t \in Hx$ we have $t = hx = xhh' \in xH$ so that $Hx \subseteq xH$. Likewise, $h^{-1}(x^{-1})^{-1}hx^{-1} = h^* \in H$ gives $xh = hh^*x$ and so if $t \in xH$ then $t = xh = hh^*x \in Hx$. Thus $xH = Hx$ as required.

1.33 N is normal in G if and only if, for all $x \in G$ and all $n \in N$ we have $x^{-1}nx \in N$, i.e. $nx \in xN$, i.e. $nxN = xN$.

Take $N = 3\mathbb{Z}$ which is a normal subgroup of the additive group $\mathbb{Z}$. Then $1 + 3 + 1 = 5 \notin N$ shows that (a) and (b) are not equivalent.

1.34 If H is of index 2 then H has only two right cosets and likewise two left cosets. Each of the coset decompositions must therefore coincide with $\{H, G \setminus H\}$. Thus if $x \notin H$ we have $xH = G \setminus H = Hx$, whereas if $x \in H$ then clearly $xH = H = Hx$. Thus we have $xH = Hx$ for every $x \in G$ so H is normal.

It is readily seen that products of elements of $A \cup B$ behave as follows:

$$XY \in \begin{cases} A & \text{if } X, Y \in A; \\ A & \text{if } X, Y \in B; \\ B & \text{if } X \in A, Y \in B; \\ B & \text{if } X \in B, Y \in A. \end{cases}$$

Thus $A \cup B$ is stable under multiplication. Since every element of $A \cup B$ is invertible, with $X \in A \Rightarrow X^{-1} \in A$ and $X \in B \Rightarrow X^{-1} \in B$, it follows that $A \cup B$ is a group.

The above observations on products also show that A is a subgroup of $A \cup B$ and that the cosets of A in $A \cup B$ are A and B. It follows by the first part of the question that A is a normal subgroup.

The set P of matrices of the form $\begin{bmatrix} a & 0 \\ 0 & 1 \end{bmatrix}$ is clearly a subgroup of A, and is a normal subgroup of A since these matrices commute with every matrix in A. However, P is not a normal subgroup of $A \cup B$; for example, if

$$X = \begin{bmatrix} 2 & 0 \\ 0 & 1 \end{bmatrix} \quad \text{and} \quad Y = \begin{bmatrix} 0 & 1 \\ 1 & 0 \end{bmatrix}$$

then we have

$$Y^{-1}XY = \begin{bmatrix} 1 & 0 \\ 0 & 2 \end{bmatrix} \notin P.$$

1.35 It is readily checked that G is a group. The identity element is the 3×3 identity matrix and the inverse of $\begin{bmatrix} 1 & a & b \\ 0 & 1 & c \\ 0 & 0 & 1 \end{bmatrix}$ is the matrix $\begin{bmatrix} 1 & -a & ac-b \\ 0 & 1 & -c \\ 0 & 0 & 1 \end{bmatrix}$.

If $h \in H$ then h commutes with every element of G; for

$$\begin{bmatrix} 1 & x & y \\ 0 & 1 & z \\ 0 & 0 & 1 \end{bmatrix}\begin{bmatrix} 1 & 0 & a \\ 0 & 1 & 0 \\ 0 & 0 & 1 \end{bmatrix} = \begin{bmatrix} 1 & x & y+a \\ 0 & 1 & z \\ 0 & 0 & 1 \end{bmatrix} = \begin{bmatrix} 1 & 0 & a \\ 0 & 1 & 0 \\ 0 & 0 & 1 \end{bmatrix}\begin{bmatrix} 1 & x & y \\ 0 & 1 & z \\ 0 & 0 & 1 \end{bmatrix}.$$

Thus $gH = Hg$ for every $g \in G$ and so H is normal.

It is easy to check that K is a subgroup of G. Now K is not normal in G. For if $x = \begin{bmatrix} 1 & 0 & 0 \\ 0 & 1 & 1 \\ 0 & 1 & 1 \end{bmatrix}$ then xK is the set of matrices of the form $\begin{bmatrix} 1 & a & 0 \\ 0 & 1 & 1 \\ 0 & 0 & 1 \end{bmatrix}$ whereas Kx is the set of matrices of the form $\begin{bmatrix} 1 & a & a \\ 0 & 1 & 1 \\ 0 & 0 & 1 \end{bmatrix}$, so that $xK \neq Kx$.

1.36 The property that there exists a positive integer n such that $(\forall x,y\in G)(xy)^n = x^n y^n$ may be expressed in the form : the mapping $f: G\to G$ described by $f(x)=x^n$ is a morphism. Consequently $G^n = \operatorname{Im} f$ is a subgroup of G which is normal since $gx^ng^{-1}=(gxg^{-1})^n\in G^n$ for every $g\in G$. Also, $G_n=\operatorname{Ker} f$ is a normal subgroup of G. Applying the first isomorphism theorem we see that

$$G^n = \operatorname{Im} f \cong G/\operatorname{Ker} f = G/G_n.$$

1.37 (*a*) (1478)(265)(39).
(*b*) (18)(364)(57), (134)(26)(587), (13478652).
(*c*) (13745).

1.38 Yes : (12)(34).(13)(24) = (14)(23) = (13)(24).(12)(34);
No : (12)(24).(13)(34) = (13)(24) but (13)(34).(12)(24) = (12)(34).

1.39 Consider a cycle $\sigma=(i_1 i_2 \dots i_r)$. Clearly we have $\sigma^{r+1}=\sigma$ and so $\sigma^r=1$. Also, for $s<r$ we have $\sigma^{s+1}\neq\sigma$ and so $\sigma^s\neq 1$. Hence σ is of order r.

Suppose now that $\sigma=\alpha\beta$ where α, β are disjoint cycles of order r, s respectively. Let $t=\operatorname{lcm}\{r,s\}$. Since disjoint cycles commute we have $\sigma^t=\alpha^t\beta^t$. Now $r|t$ and $s|t$, and $\alpha^r=1=\beta^s$; hence $\alpha^t=1=\beta^t$ and consequently $\sigma^t=1$. But t is the least integer with this property, since for $u<t$ we have $\alpha^u\beta^u=1$ implies $\alpha^u=1=\beta^u$. It follows that σ has order t.

$\sigma=(138)(27)(4965)$.
The order of σ is lcm $\{3,2,4\}=12$.
$1000=83.12+4$ so $\sigma^{1000}=\sigma^4=(138)$.

1.40 (*a*) We have that

$$(12)^{-1}(135)^{-1}(1579)(135)(12) = (12)(153)(1579)(135)(12) = (5379)$$

which is of order 4.

Likewise, $(579)^{-1}(123)(579)=(123)$ is of order 3.

(*b*) The inverse of (135)(24) is clearly (153)(24). Thus

$$\tau\sigma\tau^{-1} = (135)(24).(12)(3456).(153)(24) = (1652)(34).$$

1.41 First establish the result when $r=2$; i.e. that for a transposition (k_1k_2) we have that $\sigma(k_1k_2)\sigma^{-1}=(\sigma(k_1)\sigma(k_2))$. This is equivalent to

$$\sigma(k_1k_2)=(\sigma(k_1)\sigma(k_2))\sigma;$$

and this equality results from the observation that swapping k_1 and k_2 then applying σ (i.e. the left hand side) gives the same result as first applying σ then swapping $\sigma(k_1)$ and $\sigma(k_2)$ (i.e. the right hand side).

For the general case, use the fact that every cycle can be written as a product of transpositions:

$$(k_1k_2 \cdots k_r) = (k_1k_2)(k_2k_3) \cdots (k_{r-1}k_r).$$

Thus the result follows from the observation that

$$\begin{aligned}\sigma(k_1k_2 \cdots k_r)\sigma^{-1} &= \sigma(k_1k_2)(k_2k_3) \cdots (k_{r-1}k_r)\sigma^{-1} \\ &= \sigma(k_1k_2)\sigma^{-1} . \sigma(k_2k_3)\sigma^{-1} \cdots \sigma(k_{r-1}k_r)\sigma^{-1}.\end{aligned}$$

It is readily verified that $\{(1), (12)(34), (13)(24), (14)(23)\}$ is a subgroup of S_4. By the above, this subgroup is invariant under conjugation by $\sigma \in S_4$; hence it is a normal subgroup.

1.42 Associativity of the operation on $G_1 \times G_2$ follows from the associativity of the group operations on both G_1 and G_2. The identity element is $(1, 1)$ and the inverse of (g_1, g_2) is (g_1^{-1}, g_2^{-1}).

It is easy to verify that H_1 and H_2 are subgroups of the cartesian product group. They are in fact normal subgroups. For example, we have that

$$(x, y)^{-1}(g_1, 1)(x, y) = (x^{-1}g_1x, y^{-1}1y) = (x^{-1}g_1x, 1)$$

which shows that H_1 is normal.

(Alternatively, it could be noted that the mapping $\vartheta : G_1 \times G_2 \to G_2$ described by $\vartheta(g_1, g_2) = g_2$ is a group morphism whose kernel is H_1 which is then a normal subgroup. Similarly, H_2 is the kernel of the group morphism $\varphi : G_1 \times G_2 \to G_1$ described by $\varphi(g_1, g_2) = g_1$.)

For the last part it suffices to note that

$$(g_1, 1)(1, g_2) = (g_1, g_2) = (1, g_2)(g_1, 1).$$

1.43 (*a*) Not a morphism since $[0]_{12}$, the identity element of $\mathbb{Z}_{12}$, is not mapped to $[0]_{12}$.

(*b*) Since C_{12} is abelian we have

$$f(gh) = (gh)^3 = g^3h^3 = f(g)f(h)$$

so f is a group morphism. Writing $C_{12} = \{1, a, a^2, \ldots, a^{11}\}$ we have $\operatorname{Ker} f = \{1, a^4, a^8\}$ and $\operatorname{Im} f = \{1, a^3, a^6, a^9\}$.

(*c*) This is a group morphism since

$$\begin{aligned}f(x + y) = ([x + y]_2, [x + y]_4) &= ([x]_2 + [y]_2, [x]_4 + [y]_4) \\ &= ([x]_2, [x]_4)([y]_2, [y]_4) \\ &= f(x)f(y).\end{aligned}$$

Clearly, $\operatorname{Im} f = \{(0, 0), (1, 1), (0, 2), (1, 3)\} \simeq \mathbb{Z}_4$ and $\operatorname{Ker} f = 4\mathbb{Z}$.

(*d*) The mapping so defined is a group morphism since

$$f([x]_8 + [y]_8) = f([x+y]_8) = [x+y]_2$$
$$= [x]_2 + [y]_2 = f([x]_8) + f([y]_8).$$

$\mathrm{Im}\, f = \mathbb{Z}_2$ and $\mathrm{Ker}\, f = \{[2n]_8 \mid n \in \mathbb{Z}\}$.

(*e*) Not a morphism; for example,

$$f((1,k)(h,1)) = f(h,k) = (12)(123) = (23)$$

but $f(1,k)f(h,1) = (123)(12) = (13)$.

(*f*) This is a morphism, for $f(\pi\varphi)$ is easily seen to have the same effect on each i as $f(\pi)f(\varphi)$. $\mathrm{Im}\, f$ is the subgroup of S_{n+1} consisting of all permutations fixing $n+1$, and $\mathrm{Ker}\, f = \{1\}$.

1.44 Consider a morphism $f : \mathbb{Z} \to \mathbb{Q}$. Suppose that $f(1) = q$; then for every $n \in \mathbb{Z}$ we have $f(n) = nq$ so that f is completely determined by $f(1)$.

1.45 G consists of the matrices

$$A_n = \begin{bmatrix} 1-n & -n \\ n & 1+n \end{bmatrix}.$$

Observe that $\det A_n = 1$ so A_n is invertible. Now observe by a simple matrix multiplication that $A_m A_n = A_{m+n}$. Since $A_0 = I_2$ it follows that $A_n^{-1} = A_{-n} \in G$, whence G is a group.

Consider now the mapping $\vartheta : \mathbb{Z} \to G$ given by $\vartheta(n) = A_n$. This is a group morphism since

$$f(m+n) = A_{m+n} = A_m A_n = f(m)f(n).$$

Clearly, ϑ is a bijection, whence it is an isomorphism.

As for the set G_1 of matrices of the form

$$B_n = \begin{bmatrix} 1-2n & n \\ -4n & 1+2n \end{bmatrix}$$

we note that $\det B_n = 1$ so each B_n is invertible. Observe that $B_m B_n = B_{m+n}$ and that a similar argument to the above shows that $\mathbb{Z}$ and G_1 are isomorphic groups.

It follows that G and G_1 are isomorphic.

1.46 By definition, G consists of the 2×2 matrices with entries 0 and 1 and with determinant 1; i.e. G consists of the six matrices

$$\begin{bmatrix} 1 & 0 \\ 0 & 1 \end{bmatrix}, \begin{bmatrix} 0 & 1 \\ 1 & 0 \end{bmatrix},$$

$$\begin{bmatrix} 1 & 1 \\ 0 & 1 \end{bmatrix}, \begin{bmatrix} 1 & 0 \\ 1 & 1 \end{bmatrix}, \begin{bmatrix} 0 & 1 \\ 1 & 1 \end{bmatrix}, \begin{bmatrix} 1 & 1 \\ 1 & 0 \end{bmatrix}.$$

Writing $X_1 = [0 \;\; 1]^t$, $X_2 = [1 \;\; 0]^t$, $X_3 = [1 \;\; 1]^t$ we see that $\begin{bmatrix} 0 & 1 \\ 1 & 0 \end{bmatrix}$ sends $X_1 \to X_2$, $X_2 \to X_1$, $X_3 \to X_3$ and therefore corresponds to the permutation (12). Similarly we see that

$$\begin{bmatrix} 1 & 1 \\ 0 & 1 \end{bmatrix} \text{ corresponds to } (13);$$

$$\begin{bmatrix} 1 & 0 \\ 1 & 1 \end{bmatrix} \text{ corresponds to } (23);$$

$$\begin{bmatrix} 0 & 1 \\ 1 & 1 \end{bmatrix} \text{ corresponds to } (132);$$

$$\begin{bmatrix} 1 & 1 \\ 1 & 0 \end{bmatrix} \text{ corresponds to } (123);$$

and I_2 corresponds to the identity permutation. Clearly we have that $G \cong S_3$.

1.47 Consider $\vartheta : \mathbb{Z}[X] \to \mathbb{Q}^+$ given by

$$a_0 + a_1X + \cdots + a_nX^n \to p_0^{a_0}p_1^{a_1} \cdots p_n^{a_n}$$

where $p_0 < p_1 < \cdots < p_n$ are the first $n+1$ primes. That ϑ is a group morphism follows from the fact that if $f(X) = a_0 + a_1X + \cdots + a_nX^n$ and $g(X) = b_0 + b_1X + \cdots + b_mX^m$ then (assuming without loss of generality that $n \leqslant m$)

$$\begin{aligned} \vartheta[f(X) + g(X)] &= p_0^{a_0+b_0}p_1^{a_1+b_1} \cdots p_n^{a_n+b_n}p_{n+1}^{a_{n+1}} \cdots p_m^{b_m} \\ &= (p_0^{a_0}p_1^{a_1} \cdots p_n^{a_n})(p_0^{b_0}p_1^{b_1} \cdots p_m^{b_m}) \\ &= \vartheta[f(X)]\vartheta[g(X)]. \end{aligned}$$

Suppose now that $f(X) \in \operatorname{Ker} \vartheta$. Then $p_0^{a_0}p_1^{a_1} \cdots p_n^{a_n} = 1$. But the p_i are distinct primes so we must have $a_0 = a_1 = \cdots = a_n = 0$ whence $f(X) = 0$. Thus $\operatorname{Ker} \vartheta = \{0\}$. Also $\operatorname{Im} \vartheta = \mathbb{Q}^+$ since, given any $m/n \in \mathbb{Q}^+$ (as usual in its lowest terms), we can express each of m and n as products of powers of primes (using the fundamental theorem of arithmetic) and hence find a polynomial that maps to m/n under ϑ.

1.48 A simple matrix multiplication shows that $A^3 = 0$. Now by definition we have $A_x = I_3 + xA + \frac{1}{2}x^2A^2$ and so

$$\begin{aligned} A_xA_y &= (I_3 + xA + \tfrac{1}{2}x^2A^2)(I_3 + yA + \tfrac{1}{2}y^2A^2) \\ &= I_3 + yA + \tfrac{1}{2}y^2A^2 + xA + xyA^2 + \tfrac{1}{2}x^2A^2 \quad (\text{since } A^3 = 0) \\ &= I_3 + (x+y)A + \tfrac{1}{2}(x+y)^2A^2 \\ &= A_{x+y}, \end{aligned}$$

whence we see that $G = \{A_x \mid x \in \mathbb{R}\}$ is closed under multiplication. Now $A_0 = I_3$ is clearly the identity element of G; and $A_x A_{-x} = A_0$ shows that $A_x^{-1} = A_{-x}$. Thus G is a group, which is abelian since $A_x A_y = A_{x+y} = A_y A_x$.

Consider now the mapping $f : \mathbb{R} \to G$ given by $f(x) = A_x$. This is a group morphism since

$$f(x+y) = A_{x+y} = A_x A_y = f(x)f(y).$$

Clearly, f is surjective. Now if $x \in \operatorname{Ker} f$ we have $A_x = f(x) = I_3$ and this implies that $x = 0$ (since, for example, the (3, 3)th element of A_x is $1 + \frac{1}{2}x^2$). Thus we see that f is also injective, whence it is an isomorphism.

1.49 It is a routine matter to show that G is a group. The identity element is (0, 1) and inverses are given by

$$(a, b)^{-1} = \left(-\frac{a}{b}, \frac{1}{b}\right).$$

G is not abelian; for example, (1, 2)(2, 1) = (5, 2) whereas (2, 1)(1, 2) = (3, 2).

Use the usual subgroup criterion to show that H, K, L are subgroups of G. Of these, H and L are abelian.

$H \cap K = \{(0, b) \mid b > 0\}$. Consider the mapping $f : H \cap K \to \mathbb{R}^+$ defined by $f(0, b) = b$. This mapping is clearly a bijection. Moreover,

$$f[(0, a)(0, b)] = f(0, ab) = ab = f(0, a)f(0, b)$$

so that f is also a morphism. Hence f is an isomorphism.

(a, b) has order 2 in G if and only if

$$(0, 1) = (a, b)^2 = (a + ab, b^2).$$

For this to be the case we must have $b = \pm 1$. Now $b = 1$ gives $2a = 0$ whence $a = 0$ and we have the identity element; and if $b = -1$ then we have $(a, -1)(a, -1) = (0, 1)$ for every $a \in \mathbb{R}$. Thus we see that $\{(a, -1) \mid a \in \mathbb{R}\}$ is the subset of elements of order 2.

1.50 It is a routine matter to verify that the given operation is associative. (0, 0) is the identity element, and the inverse of (a, b) is $((-1)^{b+1}a, -b)$. Thus G is a group, which is non-abelian since, for example, (0, 1)(1, 0) = (−1, 1) and (1, 0)(0, 1) = (1, 1).

The mapping $f : G \to G$ given by $f(a, b) = (0, b)$ is a morphism since

$$f((a, b)(c, d)) = f(a + (-1)^b c, b + d) = (0, b + d),$$
$$f(a, b)f(c, d) = (0, b)(0, d) = (0, b + d).$$

The kernel of f is H and so H is a normal subgroup of G.

The image of f is K so K is a subgroup of G. However, K is not a normal subgroup since, for example

$$(1, 0)(0, 1)(-1, 0) = (1, 1)(-1, 0) = (2, 1) \notin K.$$

By the first isomorphism theorem, we have $K = \operatorname{Im} f \cong G/\operatorname{Ker} f = G/H$.

1.51 It is easy to check that the given operation is associative. The element $(e, 1)$ is the identity for G, where e is the identity in A. The inverse of (a, ϵ) is $(a^{-\epsilon}, \epsilon)$ and so G is a group.

If G is abelian we must have $ab^{\epsilon} = ba^{\delta}$ for all $a, b \in A$ and $\epsilon, \delta = \pm 1$ and it is easy to see that this means we must have $a^2 = b^2 = e$ in A. So for (a) take $A = \{e\}$ the trivial group. (Note that a cyclic group of order 2 ($\mathbb{Z}_2$) or the Klein 4-group ($\mathbb{Z}_2 \times \mathbb{Z}_2$) would also do.)

For (b) take A to be any group containing an element not of order 2; for example, $A = \mathbb{Z}_3$ or $A = \mathbb{Z}$.

It is easy to check that N is a subgroup. To establish normality we show that $n \in N$, $g \in G \Rightarrow gng^{-1} \in N$: if $n = (a, 1)$ and $g = (b, \epsilon)$ then $g^{-1} = (b^{-\epsilon}, \epsilon)$ and so the second component of gng^{-1} is $\epsilon^2 = 1$ and so $gng^{-1} \in N$.

It is clear that N is isomorphic to A under the assignment $a \to (a, 1)$. It is also clear that the mapping $f : G \to E$ described by $f(a, \epsilon) = \epsilon$ is a surjective group morphism whose kernel is N. Consequently we have that $G/N \cong E$.

For the last part, take A to be the multiplicative group of non-zero real numbers. Define $\vartheta : G \to \mathbb{R} \setminus \{0\}$ by $\vartheta(a, \epsilon) = \epsilon a$. That ϑ is a group morphism follows from

$$\vartheta[(a, \epsilon)(b, \delta)] = \vartheta(ab^{\epsilon}, \epsilon\delta) = \epsilon\delta ab^{\epsilon}.$$

For, by considering separately the cases $\epsilon = 1$ and $\epsilon = -1$, it is easy to see that (with the notation of question 1.9) $\epsilon\delta ab^{\epsilon} = (\epsilon a) * (\delta b)$. Clearly, ϑ is a bijection and hence is an isomorphism.

1.52 To prove that $(xy)^n = x^n y^n$ use induction : the equality clearly holds at the anchor point $n = 1$; and for the inductive step observe that, since G is abelian,

$$(xy)^{n+1} = x(yx)^n y = x(xy)^n y = xx^n y^n y = x^{n+1} y^{n+1}.$$

Suppose now that $x, y \in H$. Then there exist m, n such that $x^m = 1 = y^n$. Now, using the above, we have that

$$\begin{aligned}(xy^{-1})^{mn} &= x^{mn}(y^{-1})^{mn} = (x^m)^n[(y^{-1})^n]^m \\ &= (x^m)^n[(y^n)^{-1}]^m = 1\end{aligned}$$

and so $xy^{-1} \in H$. Thus H is a subgroup.

Consider now $G=\{z\in\mathbb{C}\mid |z|=1\}$. Every element of G is of the form $e^{\vartheta i}$ where $\vartheta\in\mathbb{R}$ is determined to within a multiple of 2π. Now

$$\begin{aligned}(e^{\vartheta i})^n=1 &\Leftrightarrow \cos n\vartheta+i\sin n\vartheta=1\\ &\Leftrightarrow n\vartheta=2m\pi \quad \text{for some integer} \quad m\\ &\Leftrightarrow \vartheta=\alpha\pi \quad \text{for some} \quad \alpha\in\mathbb{Q}.\end{aligned}$$

Thus $H=\{e^{\alpha\pi i}\mid \alpha\in\mathbb{Q}\}$.

We now observe that if $e^{\vartheta i}=e^{\varphi i}$ then $e^{(\vartheta-\varphi)i}=1$ and so $\vartheta-\varphi=2n\pi$ for some $n\in\mathbb{Z}$ and hence

$$\frac{\vartheta}{\pi}-\frac{\varphi}{\pi}\in\mathbb{Q}.$$

Consequently we have that

$$e^{\vartheta i}=e^{\varphi i}\Rightarrow\frac{\vartheta}{\pi}+\mathbb{Q}=\frac{\varphi}{\pi}+\mathbb{Q}$$

and so we can define a mapping $f:G\to\mathbb{R}/\mathbb{Q}$ by $f(e^{\vartheta i})=\dfrac{\vartheta}{\pi}+\mathbb{Q}$. Since

$$\begin{aligned}f(e^{\vartheta i}e^{\varphi i})=f(e^{(\vartheta+\varphi)i})&=\frac{\vartheta+\varphi}{\pi}+\mathbb{Q}\\ &=\left(\frac{\vartheta}{\pi}+\mathbb{Q}\right)+\left(\frac{\varphi}{\pi}+\mathbb{Q}\right)\\ &=f(e^{\vartheta i})+f(e^{\varphi i})\end{aligned}$$

we see that f is a group morphism.

Now f is surjective since $(\forall r\in\mathbb{R})\,f(e^{r\pi i})=r+\mathbb{Q}$.

Finally

$$\begin{aligned}\operatorname{Ker} f=\left\{e^{\vartheta i}\mid\frac{\vartheta}{\pi}+\mathbb{Q}=0+\mathbb{Q}\right\}&=\left\{e^{\vartheta i}\mid\frac{\vartheta}{\pi}\in\mathbb{Q}\right\}\\ &=\{e^{\vartheta i}\mid\vartheta=\pi\alpha\ (\alpha\in\mathbb{Q})\}\\ &=H,\end{aligned}$$

and by the first isomorphism theorem it now follows that

$$G/H=G/\operatorname{Ker} f\simeq \operatorname{Im} f=\mathbb{R}/\mathbb{Q}.$$

1.53 $H=\{(1),\ (12)(34),\ (13)(24),\ (14)(23)\}$. Now check that for every $h\in H$ we have $(123)^{-1}h(123)\in H$, so that H is normal in G.

G/H is cyclic of order 3 generated by $(123)H$.

K is normal in H since H is abelian. But $(12)(34) \in K$ and

$$(123)^{-1}(12)(34)(123) = (321)(12)(34)(123) = (13)(24) \notin K$$

so K is not normal in G.

1.54 $((25)(34))^{-1}(12345)(25)(34) = (25)(34)(12345)(25)(34) = (54321) \in H$ since $(54321) = (12345)^{-1}$. G/H is generated by $(25)(34)H$ which has order 2 and so $|G/H| = 2$. Hence $|G| = |H|.|G/H| = 5.2 = 10$.

Fig.S1.4

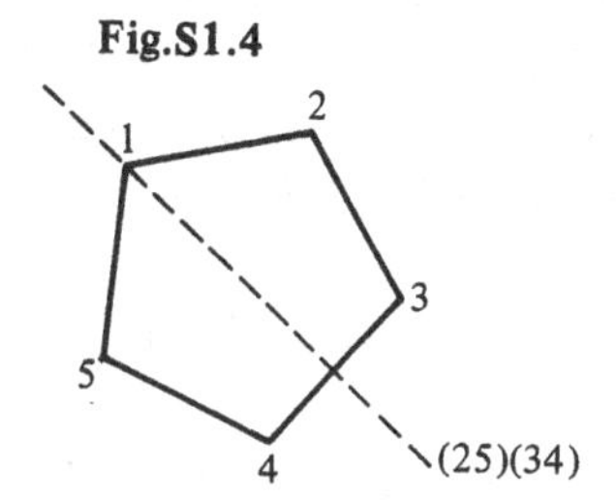

The rotational symmetries are generated by (12345). The only other symmetries are reflections in lines joining the centre of the pentagon to the mid-points of the edges (see question 1.11). Each of these can be obtained by a rotation followed by the reflection (25)(34) and the inverse rotation. Hence G is isomorphic to a group of symmetries. See Fig. S1.4.

1.55 Use the usual criterion to show that $n\mathbb{Z}$ is a subgroup of $m\mathbb{Z}$, and that $m\mathbb{Z}/n\mathbb{Z}$ is a subgroup of $\mathbb{Z}/n\mathbb{Z}$. To show that the given prescription defines a map we must show that if $a + n\mathbb{Z} = b + n\mathbb{Z}$ then $f(a + n\mathbb{Z}) = f(b + n\mathbb{Z})$. But $a + n\mathbb{Z} = b + n\mathbb{Z}$ implies $a - b \in n\mathbb{Z}$ so $a - b \in m\mathbb{Z}$ since m divides n. Hence $a + m\mathbb{Z} = b + m\mathbb{Z}$ as required.

We now observe that

$$\begin{aligned}\operatorname{Ker} f &= \{a + n\mathbb{Z} \in \mathbb{Z}/n\mathbb{Z} \mid m \text{ divides } a\} \\ &= m\mathbb{Z}/n\mathbb{Z}.\end{aligned}$$

The first isomorphism theorem now gives

$$\frac{(\mathbb{Z}/n\mathbb{Z})}{(m\mathbb{Z}/n\mathbb{Z})} \cong \frac{\mathbb{Z}}{m\mathbb{Z}}.$$

1.56 It is a routine task to show that Δ is associative (see Book 1, question 1.10). Clearly, the identity element under Δ is $\emptyset$ and $E\Delta E = \emptyset$, which shows that

$E^{-1} = E$ and hence $\mathbf{P}(X)$ is a group. It is clear that every non-identity element is of order 2.

If $X = \{1, 2\}$ then $|\mathbf{P}(X)| = 4$ and, since every non-identity element is of order 2, the group cannot be cyclic. We must therefore have $\mathbf{P}(X) \cong C_2 \times C_2$.

1.57 That $\varphi_g : aH \to gaH$ is a bijection is immediate from the observation that $aH \to g^{-1}aH$ is an inverse for it.

Now define a mapping ϑ from G to the permutation group on S (which is S_k since $|S| = k$) by $\vartheta(g) = \varphi_g$. Then $\vartheta(gh)$ is the map $aH \to ghaH$, and $\vartheta(g)\vartheta(h)$ is the map $aH \to g(haH)$. Consequently we see that ϑ is a morphism. (Note that if $H = \{1\}$ then $S \cong G$ and we obtain Cayley's theorem.)

To find Ker ϑ look for elements of G which lead to the identity permutation. In particular, the coset H maps to gH which must be H and so $g \in H$. Thus we see that Ker $\vartheta \subseteq H$.

If N is a normal subgroup of G and $N \subseteq H$ we show that $N \subseteq$ Ker ϑ. Let $n \in N$ and consider $\vartheta(n)$. This is the map $aH \to naH$. But $na = an'$ for some $n' \in N$ (since N is normal) and hence $naH = an'H = aH$ since $n' \in N \subseteq H$. Hence $\vartheta(n)$ is the identity permutation and $n \in$ Ker ϑ so that $N \subseteq$ Ker ϑ.

Suppose now that H contains no non-trivial normal subgroup of G. Then Ker $\vartheta = \{1\}$ and ϑ is injective. Hence Im $\vartheta \cong G$ and has order $|G|$. But Im ϑ is a subgroup of S_k which has order $k!$. So, by Lagrange's theorem, we have the contradiction that $|G|$ divides $k!$.

For the last part we note that since the index of H in G is 9, and 99 does not divide 9!, H contains a non-trivial normal subgroup of G. But H has prime order and so by Lagrange's theorem the only non-trivial subgroup of H is H itself. Hence H is a normal subgroup of G.

1.58 For each of the four groups we give a property which the group possesses and which the other three groups do not. This shows that no two of the groups can be isomorphic.

(*a*) Every element has order 2 (see question 1.56).

(*b*) The group is cyclic.

(*c*) The group is not abelian.

(*d*) The group is isomorphic to $C_4 \times C_2$. For, 2 generates a cyclic subgroup H of order 4, 11 generates a cyclic group K of order 2, $H \cap K = 1$ and the group is abelian.

1.59 The first part is routine.

(a) If $f: \mathbb{Z} \to \mathbb{Z}$ is an isomorphism then $f(1)$ has a multiplicative inverse in $\mathbb{Z}$ so $f(1) = \pm 1$. Hence Aut $\mathbb{Z} = \{\pm 1\}$ under multiplication and so is isomorphic to C_2.

(b) Every morphism $f: \mathbb{Z}_n \to \mathbb{Z}_n$ is determined by $f(1)$. As before, $f(1)$ must be invertible if f is to have an inverse. Using the euclidean algorithm, we see that $f(1)$ must be coprime to n. The converse is also true, so f is an isomorphism if and only if $f(1) = k$ where k is coprime to n. Thus Aut $\mathbb{Z}_n$ is isomorphic to the multiplicative group of integers that are coprime to n.

(c) If G is the Klein 4-group then it is readily seen that every bijection from G to G which takes the identity to itself is a morphism; this is because G has the property that the product of any two non-identity elements is the third non-identity element. Hence any permutation of the three non-identity elements gives an automorphism of G. Thus Aut $G \cong S_3$.

1.60 First we show that $Z(G) = \{Z^*, Z^*(-1, 1)\}$ so $Z(G)$ has order 2. It is clear that $(-1, 1)$ commutes with every element of H and so certainly we have

$$Z^*(-1, 1)(x, y) = Z^*(x, y)(-1, 1)$$

for all $(x, y) \in H$. On the other hand, suppose that $Z^*(x_1, y_1) \in Z(G)$. Then if (x_2, y_2) is any element of it we have

$$(x_1, y_1)(x_2, y_2) = (x_2, y_2)(x_1, y_1)Z$$

where $Z = (1, 1)$ or $(-1, -1)$. In other words,

$$(x_1x_2, y_1y_2) = (x_2x_1, y_2y_1) \quad \text{or} \quad (-x_2x_1, -y_2y_1).$$

Now by symmetry we need consider only elements (x_1, y_1) of the form (i, i), (i, −i), (i, j), (i, k), (i, 1). Note that (−i, −i) = (−1, −1)(i, i) and so they are in the same coset relative to H and thus represent the same element of G. We choose a specific (x_2, y_2) for each of these five elements to show that in each case $Z^*(x_1, y_1)$ is not in the centre of G.

(1) (i, i)(j, 1) = (ij, i) and (j, 1)(i, i) = (−ij, i) = (−1, 1)(ij, 1). But $(-1, 1) \notin Z^*$ and so $Z^*(\mathrm{i}, \mathrm{j})(\mathrm{j}, 1) \neq Z^*(\mathrm{j}, 1)(\mathrm{i}, \mathrm{j})$.

(2) (i, −i)(−j, 1) = (−ij, −i) and (−j, 1)(i, −i) = (−1, 1)(−ij, −i) = (ij, −i).

(3) (i, j)(j, 1) = (ij, j) and (j, 1)(i, j) = (−ij, j) = (−1, 1)(ij, j).

(4) (i, k)(k, 1) = (ik, k) and (k, 1)(i, k) = (−ik, k) = (−1, 1)(ik, k).

(5) (i, 1)(j, 1) = (ij, 1) and (j, 1)(i, 1) = (−ij, 1) = (−1, 1)(ij, 1).

This proves that $Z(G) = \{Z^*, Z^*(-1, 1)\}$.

If we take a closer look, we can see that we have proved in addition that $G/Z(G)$ is abelian. Hence, if $g_1, g_2 \in G$ we have $g_1 g_2 = g_2 g_1 z$ where $z \in Z(G)$.

Suppose that $G = G_1 \times G_2$. Then either $G_1 \cap Z(G) = 1$ or $Z(G)$ since $Z(G)$ has only two elements. Suppose without loss of generality that $G_1 \cap Z(G) = 1$. Then G_1 is abelian. For, if $x, y \in G_1$ we have $xy = yxz$ for some $z \in Z(G)$ whence $xyx^{-1}y^{-1} \in Z(G) \cap G_1 = 1$ and hence $xy = yx$. Finally, $G_1 = 1$. For, every element of G_1 commutes with every element of G_2 (since $G = G_1 \times G_2$) and hence $G_1 \leqslant Z(G)$. But $G_1 \cap Z(G) = 1$ and consequently $G_1 = 1$.

1.61 Note that

$$S = \{A \in \mathrm{Mat}_{2\times 2}(\mathrm{IR}) \mid \det(X + I_2) \neq 0\}.$$

Thus, if $A, B \in S$ we have

$$\begin{aligned} 0 &\neq \det(A + I_2)\det(B + I_2) \\ &= \det[(A + I_2)(B + I_2)] \\ &= \det(A + B + AB + I_2) \end{aligned}$$

and consequently $A + B + AB \in S$.

To prove that $*$ is associative we observe that

$$A * B = A + B + AB = (A + I_2)(B + I_2) - I_2.$$

It follows that

$$A * (B * C) = (A + I_2)(B + I_2)(C + I_2) - I_2 = (A * B) * C.$$

Clearly, $A * 0 = A$ and $0 * B = B$ and so 0 is the identity element for $*$. Now

$$\begin{aligned} A * B = 0 \Leftrightarrow A + B + AB = 0 &\Leftrightarrow (A + I_2)B = -A \\ &\Leftrightarrow B = -(A + I_2)^{-1}A \end{aligned}$$

so the inverse of A with respect to $*$ is $-(A + I_2)^{-1}A$. Thus S is a group.

Consider now the mapping $f: S \to M$ given by $f(A) = A + I_2$. That f is a morphism follows from the observation

$$f(A * B) = (A * B) + I_2 = (A + I_2)(B + I_2) = f(A)f(B).$$

If $f(A) = f(B)$ then $A + I_2 = B + I_2$ gives $A = B$, so f is injective. Finally, if $X \in M$ then clearly $X - I_2 \in S$ and $f(X - I_2) = X$, so that f is also surjective. Thus f is a group isomorphism.

1.62 We have that

$$\left(\frac{x+y}{1+xy}\right)^2 < 1 \Leftrightarrow x^2 + 2xy + y^2 < 1 + 2xy + x^2y^2$$
$$\Leftrightarrow 1 + x^2y^2 - x^2 - y^2 > 0$$
$$\Leftrightarrow (1-x^2)(1-y^2) > 0.$$

Consequently, if $x, y \in G$ then

$$\frac{x+y}{1+xy} \in G.$$

For $x * y = \dfrac{x+y}{1+xy}$ it is readily seen that

$$x * (y * z) = \frac{x+y+z+xyz}{1+xy+xz+yz} = (x * y) * z.$$

Thus $*$ is associative. Clearly, $*$ is commutative; and since $x * 0 = x$ it follows that 0 is the identity element. Finally, $x * (-x) = 0$ shows that $-x$ is the inverse of x under $*$. Hence G is a group.

With $f(x) = \log \dfrac{1+x}{1-x}$ we have

$$f(x * y) = \log \frac{1 + \dfrac{x+y}{1+xy}}{1 - \dfrac{x+y}{1+xy}} = \log \frac{1+xy+x+y}{1+xy-x-y}$$

$$f(x) + f(y) = \log \left(\frac{1+x}{1-x} \cdot \frac{1+y}{1-y}\right) = \log \frac{1+xy+x+y}{1+xy-x-y}$$

and so f is a morphism.

That f is injective follows from the fact that

$$f(x) = 0 \Rightarrow \log \frac{1+x}{1-x} = 0 \Rightarrow \frac{1+x}{1-x} = 1 \Rightarrow x = 0.$$

To see that f is also surjective, we observe that $x \to \dfrac{1+x}{1-x}$ maps $]-1, 1[$ onto $\mathbb{R}_+$, and $x \to \log x$ maps $\mathbb{R}_+$ onto $\mathbb{R}$. Hence f is an isomorphism.

1.63 Suppose first that there is a law of composition $*$ on H making H a group with f an isomorphism. Then $f^{-1} : H \to G$ is also an isomorphism and hence

$(\forall x, y \in H) f^{-1}(x * y) = f^{-1}(x) f^{-1}(y)$ whence we have that

$$(\forall x, y \in H) \quad x * y = f[f^{-1}(x) f^{-1}(y)].$$

Thus, if such a law $*$ exists then it must be given by the above formula, and hence is unique.

We must now show that such a law exists. The way to do so is clear: define on H the law of composition $*$ by

$$x * y = f[f^{-1}(x) f^{-1}(y)].$$

Then $*$ is associative; for

$$x * (y * z) = f[f^{-1}(x) f^{-1}(y) f^{-1}(z)] = (x * y) * z.$$

For every $x \in H$ we have $x * f(1) = f[f^{-1}(x) \cdot 1] = x$ and similarly $f(1) * x = x$. Thus $f(1)$ is the identity of H. Given $x \in H$, consider now $f[(f^{-1}(x))^{-1}] \in H$. We have

$$x * f[(f^{-1}(x))^{-1}] = f[f^{-1}(x) \cdot (f^{-1}(x))^{-1}] = f(1).$$

Thus $f[(f^{-1}(x))^{-1}]$ is the inverse of x in H and so H is a group. The bijection f is also an isomorphism since, for all $x, y \in G$, we have

$$f(x) * f(y) = f[f^{-1}(f(x)) \cdot f^{-1}(f(y))] = f(x \cdot y).$$

Solutions to Chapter 2

2.1 (*a*) To show that S is a ring under $+$ and $\cdot$ we check the appropriate axioms:

(i) $f+g$ is a mapping from $\mathbb{R}$ to $\mathbb{R}$;

(ii) For every $x \in \mathbb{R}$ we have $[(f+g)+h](x) = [f+(g+h)](x)$ and so $(f+g)+h = f+(g+h)$;

(iii) Define $0 : \mathbb{R} \to \mathbb{R}$ by $0(x) = 0$ for every $x \in \mathbb{R}$. Then $0+f = f+0 = f$ for all maps $f : \mathbb{R} \to \mathbb{R}$;

(iv) Given $f : \mathbb{R} \to \mathbb{R}$ define $-f : \mathbb{R} \to \mathbb{R}$ by $(-f)(x) = -f(x)$. Then $f+(-f) = (-f)+f = 0$;

(v) $(f+g)(x) = (g+f)(x)$ for every $x \in \mathbb{R}$ so $f+g = g+f$;

(vi) $f \cdot g$ is a mapping from $\mathbb{R}$ to $\mathbb{R}$;

(vii) $[(f \cdot g) \cdot h](x) = [f \cdot (g \cdot h)](x)$ for every $x \in \mathbb{R}$ and hence $(f \cdot g) \cdot h = f \cdot (g \cdot h)$;

(viii) $[(f+g) \cdot h](x) = [f \cdot h + g \cdot h](x)$ for every $x \in \mathbb{R}$ and so $(f+g) \cdot h = f \cdot h + g \cdot h$. The other distributive law is similar.

It is clear that $f \cdot g = g \cdot f$ and so S is commutative. Also, the map $1 : \mathbb{R} \to \mathbb{R}$ given by $1(x) = 1$ for every $x \in \mathbb{R}$ is the identity of S.

S is not a ring under $+$ and $\circ$; for example, the distributive laws fail: take $g(x) = x$, $h(x) = -x$ and $f(x) = x^2$ to obtain $[f \circ (g+h)](1) = 0$ and $[f \circ g + f \circ h](1) = 2$.

(*b*) A simple check of axioms shows as in (*a*) above that S is a ring which is clearly commutative and has an identity element, namely $(1, 1)$. Now $(a, b) \in S$ has a multiplicative inverse if and only if there exists $(c, d) \in S$ such that $(a, b) \cdot (c, d) = (1, 1)$, i.e. if and only if $ac = 1$ and $bd = 1$. Thus we see that (a, b) has a multiplicative inverse if and only if $a \neq 0$ and $b \neq 0$.

2.2 (*a*) To show that $R(\lambda)$ is a ring it suffices to observe that if

$$A = \begin{bmatrix} x_1 + y_1 & y_1 \\ \lambda y_1 & x_1 \end{bmatrix}, \quad B = \begin{bmatrix} x_2 + y_2 & y_2 \\ \lambda y_2 & x_2 \end{bmatrix}$$

then $A - B \in R(\lambda)$ and $AB \in R(\lambda)$; then $R(\lambda)$ is a subring of the ring M. The identity element of $R(\lambda)$ is clearly $\begin{bmatrix} 1 & 0 \\ 0 & 1 \end{bmatrix}$. Suppose that $\begin{bmatrix} x+y & y \\ -y & x \end{bmatrix}$ is a non-zero element of $R(-1)$ (so that x and y are not both zero). Then it is readily seen that if $a = x^2 + xy + y^2$ then $a \neq 0$ and this matrix is invertible with inverse

$$\begin{bmatrix} \left(\frac{x}{a} + \frac{y}{a}\right) - \frac{y}{a} & -\frac{y}{a} \\ (-1)\left(-\frac{y}{a}\right) & \frac{x}{a} + \frac{y}{a} \end{bmatrix}$$

which belongs to $R(-1)$.

In $R(0)$ the matrix $\begin{bmatrix} 1 & 1 \\ 0 & 0 \end{bmatrix}$ has no multiplicative inverse.

(*b*) To show that F is a subring of M observe that

$$\begin{bmatrix} x & x \\ x & x \end{bmatrix} - \begin{bmatrix} y & y \\ y & y \end{bmatrix} = \begin{bmatrix} x-y & x-y \\ x-y & x-y \end{bmatrix} \in F$$

and that

$$\begin{bmatrix} x & x \\ x & x \end{bmatrix} \begin{bmatrix} y & y \\ y & y \end{bmatrix} = \begin{bmatrix} 2xy & 2xy \\ 2xy & 2xy \end{bmatrix} \in F.$$

The identity element of F is $\begin{bmatrix} \frac{1}{2} & \frac{1}{2} \\ \frac{1}{2} & \frac{1}{2} \end{bmatrix}$. If $x \neq 0$ then $\begin{bmatrix} x & x \\ x & x \end{bmatrix}$ has multiplicative inverse $\begin{bmatrix} \frac{1}{4x} & \frac{1}{4x} \\ \frac{1}{4x} & \frac{1}{4x} \end{bmatrix}$. The last part is explained by the fact that in F the identity element is not the 2×2 identity matrix I_2 of M.

2.3 Suppose that $x \in \mathbb{Z}_n$ is not invertible and that $x \neq 0$. Then $1 < x < n$ with hcf $\{x, n\} \neq 1$. Thus there exists t with $1 < t < n$ which divides both x and n, say $x = at$ and $n = bt$ where $1 < a < n$ and $1 < b < n$. Now

$$xb = atb = an \equiv 0 \pmod{n}$$

so we have $xb = 0$ in $\mathbb{Z}_n$, whence x is a zero divisor.

2.4 That K is a subring is routine. That it is commutative follows from the observation that

$$\begin{bmatrix} a & -b \\ b & a \end{bmatrix}\begin{bmatrix} c & -d \\ d & c \end{bmatrix} = \begin{bmatrix} ac-bd & -(bc+ad) \\ bc+ad & ac-bd \end{bmatrix}$$

$$= \begin{bmatrix} c & -d \\ d & c \end{bmatrix}\begin{bmatrix} a & -b \\ b & a \end{bmatrix}$$

and clearly $\begin{bmatrix} 1 & 0 \\ 0 & 1 \end{bmatrix}$ is the identity of K.

Note that 0 and 1 are the only squares in $\mathbb{Z}_3$ so for $a, b \in \mathbb{Z}_3$ we have $a^2 + b^2 = 0$ only if $a = b = 0$. Hence a non-zero element $\begin{bmatrix} a & -b \\ b & a \end{bmatrix}$ of K has inverse $\begin{bmatrix} at & bt \\ -bt & at \end{bmatrix}$ in K where $t = (a^2 + b^2)^{-1}$. Consequently, K is a field.

There are three possibilities for each of a and b, so there are $3 \times 3 = 9$ elements in K. Hence K contains eight non-zero elements. It is a routine matter to show that each of these is a power of $\begin{bmatrix} 1 & 2 \\ 2 & 1 \end{bmatrix}$, whence the result follows.

2.5 In (a), (c) and (d) we have examples of integral domains, so that $x(x+1) = 0$ implies $x = 0$ or $x + 1 = 0$ whence $x = 0$ or $x = -1$. Hence there are two solutions. In (b) it is readily seen that 0, 2, 3, 5 are solutions, so there are four solutions in $\mathbb{Z}_6$.

2.6 (a) $a = 3$ works : powers 3, 9, 27, 13, 5, 15, 11, 33, 31, 25, 7, 21, 29, 19, 23, 1.

(b) $0^2 = 0$, $1^2 = 1$, $17^2 = 17$ and $18^2 = 18$.

2.7 That E is a ring is routine. If $(a, b) \in E$ is a zero divisor then there exists $(c, d) \in E$ with $(c, d) \neq (0, 0)$ and $(a, b) \otimes (c, d) = (0, 0)$. Now $ac + bd = bc + ad = 0$ so $a^2c + abd = 0$ and $b^2c + bad = 0$ whence $a^2c = b^2c$, so $c(a^2 - b^2) = 0$. If $c = 0$ then from $ac + bd = bc + ad$ we obtain $bd = ad$ so if R has no zero divisors $a = b$ (note that $d \neq 0$ since $c = 0$ and $(c, d) \neq (0, 0)$). If $c \neq 0$ then, again using the fact that R has no zero divisors, we deduce from $c(a^2 - b^2) = 0$

that $a^2 - b^2 = 0$ and so $(a-b)(a+b) = 0$ which gives $a = b$ or $a = -b$. Thus we have that (a, a) and $(a, -a)$ are the only zero divisors in E. Moreover, E contains the zero divisors (a, a) and $(a, -a)$ whatever the properties of R since $(a, a) \otimes (a, -a) = (0, 0)$.

2.8 We have that

$$x^2 = 1 \Leftrightarrow x^2 - 1 = 0 \Leftrightarrow (x+1)(x-1) = 0$$

and since D is an integral domain there are only two solutions, given by $x + 1 = 0$ and $x - 1 = 0$.

To prove that if p is a prime then $(p-1)! \equiv -1 \pmod p$ we observe first that if $p = 2$ then the result is trivially true. Thus we can assume that $p > 2$, so that p is *odd*. Now $\mathbb{Z}_p$ is a field, and every field is an integral domain; so there are only two solutions of $x^2 = 1$ in $\mathbb{Z}_p$. Clearly, these are 1 and $p-1$. Consider now the product

$$1 . 2 . 3 \ldots (p-1)$$

of all the non-zero elements of $\mathbb{Z}_p$. In this we can pair off every element with its inverse and thereby see that the product reduces to $p-1$; for, $1^{-1} = 1$, the number of factors is *even*, and $p-1$ is the only element of order 2. Consequently it follows that $(p-1)! \equiv -1 \pmod p$.

2.9 First, note that $*$ is associative; this follows from the fact that

$$(a * b) * c = a + b + c - ab - ac - bc + abc = a * (b * c).$$

Now G is closed under $*$ since if $a, b \in G$ then there exist $u, w \in E$ with $a * u = 0$ and $b * w = 0$. Then, using the fact that $*$ is associative, we see that

$$\begin{aligned}(a * b) * (w * u) &= a * b * w * u \\ &= a * 0 * u \\ &= a * u \\ &= 0.\end{aligned}$$

Similarly, $(w * u) * (a * b) = 0$ and hence $a * b \in G$. That G is a group under $*$ now follows from the fact that $a * 0 = a = 0 * a$, so that 0 is the identity element, and inverses exist by the definition of G.

If $e - a \in E$ is invertible then there exists x, which can be written in the form $e - b$, such that $e = (e-a)x = (e-a)(e-b)$. Expanding the right hand side, we obtain from this $a * b = a + b - ab = 0$. Likewise, the equation $(e-b)(e-a) = e$ gives $b * a = 0$ and hence $a \in G$.

Conversely, if $a \in G$ then there exists $b \in G$ such that $a * b = 0$, whence $a + b - ab = 0$ and so $(e-a)(e-b) = e$, so that $e - a$ is invertible in E.

2.10 That H is a subring of $\mathrm{Mat}_{2\times 2}(\mathbb{C})$ is routine. A non-zero element of H has an inverse in H if and only if it has an inverse in $\mathrm{Mat}_{2\times 2}(\mathbb{C})$ and this inverse belongs to H. Now we have that the inverse in $\mathrm{Mat}_{2\times 2}(\mathbb{C})$ of

$$\begin{bmatrix} a+bi & c+di \\ -c+di & a-bi \end{bmatrix}$$

is given by

$$\alpha^{-1}\begin{bmatrix} a-bi & -c-di \\ c-di & a+bi \end{bmatrix}$$

where $\alpha = a^2+b^2+c^2+d^2 \neq 0$. Note that $\alpha = 0$ only if $a=b=c=d=0$. This inverse is of the form

$$\begin{bmatrix} a^*+b^*i & c^*+d^*i \\ -c^*+d^*i & a^*-b^*i \end{bmatrix}$$

where $a^* = a/\alpha$, $b^* = -b/\alpha$, $c^* = -c/\alpha$, $d^* = -d/\alpha$ and therefore belongs to H. Moreover, it is clear that every non-zero element of H has an inverse, whence H has no zero divisors. But H is not a field since it fails to be commutative; for example, the matrices $\begin{bmatrix} 0 & 1 \\ -1 & 0 \end{bmatrix}$ and $\begin{bmatrix} 0 & i \\ i & 0 \end{bmatrix}$ do not commute.

If $X^2 = -I_2$ then $\det X = \pm 1$ and so $a^2+b^2+c^2+d^2 = \pm 1$. The only solutions for $a,b,c,d \in \mathbb{Z}$ are

$$\begin{aligned} a &= \pm 1, \quad b=c=d=0; \\ b &= \pm 1, \quad a=c=d=0; \\ c &= \pm 1, \quad a=b=d=0; \\ d &= \pm 1, \quad a=b=c=0. \end{aligned}$$

We can eliminate the possibility $a=\pm 1$, $b=c=d=0$ since this gives $X = \pm I_2$ so that $X^2 = I_2$. The remaining six solutions provide matrices that satisfy $X^2 = -I_2$, namely

$$\begin{bmatrix} -i & 0 \\ 0 & i \end{bmatrix}, \quad \begin{bmatrix} i & 0 \\ 0 & -i \end{bmatrix}, \quad \begin{bmatrix} 0 & -1 \\ 1 & 0 \end{bmatrix},$$

$$\begin{bmatrix} 0 & 1 \\ -1 & 0 \end{bmatrix}, \quad \begin{bmatrix} 0 & i \\ i & 0 \end{bmatrix}, \quad \begin{bmatrix} 0 & -i \\ -i & 0 \end{bmatrix}.$$

2.11 That $f(\alpha\beta) = f(\alpha)f(\beta)$ is a straightforward calculation. If α has a multiplicative inverse β then from $\alpha\beta = 1$ we obtain

$$f(\alpha)f(\beta) = f(\alpha\beta) = f(1) = 1$$

and hence $f(\alpha) = f(\beta) = 1$. Now clearly $a^2 + b^2 = 1$ implies $a = \pm 1, b = 0$ or $a = 0,\ b = \pm 1$, so 1, −1, i, −i are the only elements with multiplicative inverses.

If $f(\alpha)$ is prime then $\alpha = \beta\gamma$ gives $f(\alpha) = f(\beta\gamma) = f(\beta)f(\gamma)$ so either $f(\beta) = 1$ or $f(\gamma) = 1$. Hence either β or γ has a multiplicative inverse.

Now there is no element α with $f(\alpha) = 3$ since $a^2 + b^2 = 3$ is impossible. Thus $3 \in \mathbb{Z}[i]$ is irreducible; for $f(3) = 9$ and $3 = \alpha\beta$ would require $f(3) = f(\alpha)f(\beta)$, so if $f(\alpha) \neq 1 \neq f(\beta)$ then $f(\alpha) = f(\beta) = 3$.

$f(43i - 19) = 2210 = 2.5.13.17$.

$43i - 19 = (1 + i)(2 + i)(2 + 3i)(4 - i)$ is a product of irreducibles (since each of the factors α has $f(\alpha)$ a prime).

2.12 It is readily seen that $\sim$ is reflexive and symmetric. To show that it is also transitive, suppose that we have $(r_1s_2 - r_2s_1)t = 0$ and $(r_2s_3 - r_3s_2)t' = 0$. Multiply the first of these by $t's_3$, the second by ts_1, then add; the result is $(r_1s_3 - r_3s_1)tt's_2 = 0$ where $tt's_2 \in S$. Thus $\sim$ is also transitive.

That R_S is a commutative ring with an identity is routine. (It may help if you write r/s instead of $[(r, s)]$, because then the sums look very much like the usual sums you get in fractions.)

(*a*) The mapping described by $[(r, s)] \to r/s$ is an isomorphism.

(*b*) Writing $[(r, s)]$ as r/s we have that if $r/s \neq 0$ then $r \neq 0$ and so $s/r \in R_S$ and is the inverse of r/s.

(*c*) When $R = \mathbb{Z}_6$ and $S = \{2, 4\}$ the set M of ordered pairs has 12 elements. However, the equivalence classes under $\sim$ are

$$\{(0, 2), (0, 4), (3, 2), (3, 4)\}$$
$$\{(1, 2), (2, 4), (4, 2), (5, 4)\}$$
$$\{(4, 4), (2, 2), (1, 4), (5, 2)\}$$

and so R_S has order 3.

2.13 Note that $p^2 - pq + q^2 = (p - q)^2 + pq$. If this expression is 0 then we have, on the one hand, $pq = p^2 + q^2 \geqslant 0$ and, on the other hand, $pq = -(p - q)^2 \leqslant 0$. Consequently $pq = 0$ and hence $p^2 + q^2 = 0$ so that $p = q = 0$.

S consists of the matrices of the form $\begin{bmatrix} q & p \\ -p & q - p \end{bmatrix}$ where $p, q \in \mathbb{Q}$. It is easy to check that S is closed under subtraction and multiplication, hence is a subring of $\mathrm{Mat}_{2\times 2}(\mathbb{R})$. That S is a field follows from the fact that it is commutative and, by the above observation,

$$\det\begin{bmatrix} q & p \\ -p & q-p \end{bmatrix} = p^2 - pq + q^2$$

is 0 if and only if $p = q = 0$, so every non-zero element of S has an inverse under multiplication which is in S.

2.14 Suppose that $a + b\sqrt{p} = 0$. If $a \neq 0$ then clearly $b \neq 0$ and then $\sqrt{p} = -ab^{-1} \in F$, a contradiction. Thus $a + b\sqrt{p} = 0$ implies $a = b = 0$. The converse implication is obvious. It now follows that

$$\begin{aligned} a + b\sqrt{p} = c + d\sqrt{p} &\Leftrightarrow (a-c) + (b-d)\sqrt{p} = 0 \\ &\Leftrightarrow a - c = 0 = b - d \\ &\Leftrightarrow a = c, b = d. \end{aligned}$$

If $F[\sqrt{p}] = \{a + b\sqrt{p} \mid a, b \in F\}$ then

$$a + b\sqrt{p} - (c + d\sqrt{p}) = (a-c) + (b-d)\sqrt{p} \in F[\sqrt{p}],$$
$$(a + b\sqrt{p})(c + d\sqrt{p}) = (ac + bdp) + (ad + bc)\sqrt{p} \in F[\sqrt{p}],$$

and so $F[\sqrt{p}]$ is a subring of $\mathbb{R}$. The identity element of this ring is $1 + 0\sqrt{p}$; and since, for $a + b\sqrt{p} \neq 0$, we have

$$(a + b\sqrt{p})\left(\frac{a}{a^2 - b^2 p} - \frac{b}{a^2 - b^2 p}\sqrt{p}\right) = 1$$

with $a^2 - b^2 p \neq 0$ (otherwise $\sqrt{p} = |a/b| \in F$, a contradiction), it follows that every non-zero element of $F[\sqrt{p}]$ has an inverse. Thus $F[\sqrt{p}]$ is a field.

If $\sqrt{m} \in \mathbb{Q}[\sqrt{n}]$, say $\sqrt{m} = r + s\sqrt{n}$ where $r, s \in \mathbb{Q}$, then we have $m = (r + s\sqrt{n})^2 = r^2 + s^2 n + 2rs\sqrt{n}$. Since m is a positive integer it follows that $r \neq 0$ and $s \neq 0$, whence the contradiction $\sqrt{n} = (m - r^2 - s^2 n)/2rs \in \mathbb{Q}$. Thus we deduce that $\sqrt{m} \notin \mathbb{Q}[\sqrt{n}]$.

For the last part we observe that

$$\begin{aligned} S &= \{a + b\sqrt{m} + c\sqrt{n} + d\sqrt{(mn)} \mid a, b, c, d \in \mathbb{Q}\} \\ &= \{a + c\sqrt{n} + (b + d\sqrt{n})\sqrt{m} \mid a, b, c, d \in \mathbb{Q}\} \\ &= \{p + q\sqrt{m} \mid p, q \in \mathbb{Q}[\sqrt{n}]\} \\ &= F[\sqrt{m}] \quad \text{where} \quad F = \mathbb{Q}[\sqrt{n}], \end{aligned}$$

so S is a field.

2.15 That R is a ring is routine. To check that S is an ideal, let $f, g \in S$. Then

$$(f - g)(0) = f(0) - g(0) = 0$$

so $f-g \in S$. Also, for every $r \in R$,

$$(rf)(0) = r(0)f(0) = r(0).0 = 0,$$
$$(fr)(0) = f(0)r(0) = 0.r(0) = 0$$

so $rf \in S$ and $fr \in S$. Hence S is an ideal of R.

T is a subring of R since if $f, g \in T$ then

$$\mathrm{D}(f-g)(0) = (\mathrm{D}f - \mathrm{D}g)(0) = \mathrm{D}f(0) - \mathrm{D}g(0) = 0,$$
$$\mathrm{D}(fg)(0) = (\mathrm{D}f.g + f.\mathrm{D}g)(0) = \mathrm{D}f(0)g(0) + f(0)\mathrm{D}g(0) = 0.$$

However, T is not an ideal of R since, for example, $f: \mathbb{R} \to \mathbb{R}$ given by $(\forall x \in \mathbb{R}) f(x) = 1$ satisfies $\mathrm{D}f(0) = 0$, and so $f \in T$, but $g: \mathbb{R} \to \mathbb{R}$ given by $(\forall x \in \mathbb{R}) g(x) = x$ is in R and $fg = g \notin T$.

For the last part we note that since S and T are subrings so also is $S \cap T$. Also, if $f \in S \cap T$ then $\mathrm{D}f(0) = 0$ and $f(0) = 0$ and so, for every $g \in R$,

$$\mathrm{D}(fg)(0) = (\mathrm{D}f.g + f.\mathrm{D}g)(0) = \mathrm{D}f(0)g(0) + f(0)\mathrm{D}g(0) = 0.$$

2.16 Suppose that $a, b \in A(X)$. Then $ax = bx = 0$ for all $x \in X$ and so $(a-b)x = 0$ for all $x \in X$, whence $a - b \in A(X)$. Also, if $r \in R$ then $(ra)x = r(ax) = 0$ for all $x \in X$ and so $ra \in A(X)$. Thus $A(X)$ is a left ideal.

Suppose now that $c, d \in B(X)$. Then $xc = xd = 0$ for all $x \in X$ and so $x(c-d) = 0$ for all $x \in X$, whence $c - d \in B(X)$. Also, if $r \in R$ then $x(cs) = (xc)s = 0$ for all $x \in X$ and so $cs \in B(X)$. Thus $B(X)$ is a right ideal.

Suppose now that X is a left ideal. Let $a \in A(X)$ and $r \in R$. Then for every $x \in X$ we have $(ar)x = a(rx) = 0$ since $rx \in X$. Thus $ar \in A(X)$ and $A(X)$ is also a right ideal.

For the second part of the question we begin by noting that the given set X of matrices is closed under subtraction. If now $\begin{bmatrix} a & b \\ c & d \end{bmatrix} \in R$ then

$$\begin{bmatrix} a & b \\ c & d \end{bmatrix}\begin{bmatrix} x & 0 \\ y & 0 \end{bmatrix} = \begin{bmatrix} ax+by & 0 \\ cx+dy & 0 \end{bmatrix} \in X$$

and so X is a left ideal.

To calculate $A(X)$: for all x, y we must have

$$\begin{bmatrix} a & b \\ c & d \end{bmatrix}\begin{bmatrix} x & 0 \\ y & 0 \end{bmatrix} = \begin{bmatrix} 0 & 0 \\ 0 & 0 \end{bmatrix}$$

i.e., $ax + by = 0 = cx + dy$. Taking $x = 1, y = 0$ then $x = 0, y = 1$ we obtain $a = b = c = d = 0$. Thus $A(X)$ consists only of the zero matrix.

To calculate $B(X)$: for all x, y we must have

$$\begin{bmatrix} x & 0 \\ y & 0 \end{bmatrix}\begin{bmatrix} a & b \\ c & d \end{bmatrix} = \begin{bmatrix} 0 & 0 \\ 0 & 0 \end{bmatrix}$$

i.e., $ax = bx = ay = by = 0$, which gives $a = b = 0$. Thus $B(X)$ consists of the matrices of the form

$$\begin{bmatrix} 0 & 0 \\ c & d \end{bmatrix}$$

where $c, d \in \mathbb{R}$.

Note that $B(X)$ is not a left ideal since, for example,

$$\begin{bmatrix} 0 & 1 \\ 0 & 0 \end{bmatrix}\begin{bmatrix} 0 & 0 \\ 1 & 1 \end{bmatrix} = \begin{bmatrix} 1 & 1 \\ 0 & 0 \end{bmatrix} \notin B(X).$$

2.17 It is given that $\sqrt{I} = \{x \in R \mid (\exists n \geqslant 1)\, x^n \in I\}$. Given $x, y \in \sqrt{I}$, suppose that $x^n \in I$ and $y^m \in I$. Consider $(x-y)^{n+m}$: every term in its binomial expansion contains either x^n or y^m as a factor. Thus, since I is an ideal, we have $(x-y)^{n+m} \in I$ whence $x - y \in \sqrt{I}$. Also, if $r \in R$ then, since R is commutative, $(rx)^n = r^n x^n \in I$ whence $rx \in \sqrt{I}$. Thus $\sqrt{I}$ is an ideal of R.

If $x \in I$ then $x = x^1 \in I$ gives $x \in \sqrt{I}$; so $I \subseteq \sqrt{I}$.

By the above, we have that $\sqrt{I} \subseteq \sqrt{(\sqrt{I})}$. To obtain the reverse inclusion, suppose that $x \in \sqrt{(\sqrt{I})}$. Then we have $x^n \in \sqrt{I}$ for some n, whence $x^{nm} \in I$ for some n, m. Consequently we see that $x \in \sqrt{I}$.

We now note that $I \subseteq J \Rightarrow \sqrt{I} \subseteq \sqrt{J}$. In fact, if $x \in \sqrt{I}$ then, for some $n, x^n \in I \subseteq J$ gives $x \in \sqrt{J}$. To establish (*a*) we observe that

$$\begin{aligned} I \subseteq J \subseteq \sqrt{I} &\Rightarrow \sqrt{I} \subseteq \sqrt{J} \subseteq \sqrt{(\sqrt{I})} = \sqrt{I} \\ &\Rightarrow \sqrt{J} = \sqrt{I}. \end{aligned}$$

As for (*b*), it is clear from the above observation that $\sqrt{(I \cap J)} \subseteq I$ and $\sqrt{(I \cap J)} \subseteq \sqrt{J}$ and so $\sqrt{(I \cap J)} \subseteq \sqrt{I} \cap \sqrt{J}$. To obtain the reverse inclusion, let $x \in \sqrt{I} \cap \sqrt{J}$. Then there exist n, m such that $x^n \in I$ and $x^m \in J$. Since I and J are ideals it follows that $x^{n+m} = x^n x^m \in I \cap J$, whence $x \in \sqrt{(I \cap J)}$.

2.18 That R is a subring of $\text{Mat}_{2\times 2}(\mathbb{R})$ is routine. Also, simple matrix multiplication will show that R is commutative. The identity element of R is the 2×2 identity matrix I_2 of $\text{Mat}_{2\times 2}(\mathbb{R})$.

Again, simple computations show that if $A, B \in I$ then $A - B \in I$, and if $X \in R$ then $XA \in I$. Thus we see that I is an ideal of R.

Taking $x = 1$, $y = 0$ we see that $\begin{bmatrix} 1 & \sqrt{5} \\ -\sqrt{5} & 1 \end{bmatrix} \in I$; and taking $x = 3$,

$y = -1$ we see that $\begin{bmatrix} 3 & 0 \\ 0 & 3 \end{bmatrix} \in I.$

Suppose, by way of obtaining a contradiction, that $I = (r)$. Then for some $t, u \in R$ we have

$$\begin{bmatrix} 1 & \sqrt{5} \\ -\sqrt{5} & 1 \end{bmatrix} = rt, \quad \begin{bmatrix} 3 & 0 \\ 0 & 3 \end{bmatrix} = ru.$$

Then $6 = \det r \det t$ and $9 = \det r \det u$ give $\det r \in \{1, 3\}$, since for $r = \begin{bmatrix} a & b\sqrt{5} \\ -b\sqrt{5} & a \end{bmatrix} \in R$ we have $\det r = a^2 + 5b^2$ which is an integer. Now $a^2 + 5b^2 = 3$ is impossible; and $a^2 + 5b^2 = 1$ implies that r is invertible, whence the contradiction $I = (r) = R$.

2.19 (*a*) False. See the example in (*c*) below.

(*b*) True. Let A, B be left ideals with $a_1, a_2 \in A$ and $b_1, b_2 \in B$. Then we have

$$(a_1 + b_1) - (a_2 + b_2) = (a_1 - a_2) + (b_1 - b_2) \in A + B,$$

and, for any $x \in R$,

$$x(a + b) = xa + xb \in A + B.$$

(*c*) False. Take A to be the set of matrices of the form $\begin{bmatrix} a & 0 \\ b & 0 \end{bmatrix}$ where $a, b \in \mathbb{R}$, and B to be the set of matrices of the form $\begin{bmatrix} a & b \\ 0 & 0 \end{bmatrix}$ where $a, b \in \mathbb{R}$. Then A is a left ideal and B is a right ideal of $\mathrm{Mat}_{2\times 2}(\mathbb{R})$. But $A + B$ consists of the matrices of the form $\begin{bmatrix} a & b \\ c & 0 \end{bmatrix}$ where $a, b, c \in \mathbb{R}$, and this is not a subring of $\mathrm{Mat}_{2\times 2}(\mathbb{R})$ since, for example,

$$\begin{bmatrix} 0 & 0 \\ 1 & 0 \end{bmatrix} \begin{bmatrix} 0 & 1 \\ 0 & 0 \end{bmatrix} = \begin{bmatrix} 0 & 0 \\ 0 & 1 \end{bmatrix} \notin A + B.$$

(*d*) True. See (*b*) above.

2.20 That $\mathrm{Hom}(A)$ is a ring is a routine verification of the axioms. Note that in this case $f \circ (g + h) = (f \circ g) + (f \circ h)$ follows from the fact that f is a group morphism.

If $A = \mathbb{Z}$ then every morphism $f : \mathbb{Z} \to \mathbb{Z}$ is determined by $f(1)$, since $f(n) = f(1 + 1 + \cdots + 1) = f(1) + \cdots + f(1) = nf(1)$. The mapping ϑ : $\mathrm{Hom}(\mathbb{Z}) \to \mathbb{Z}$ described by $\vartheta(f) = f(1)$ is then a ring isomorphism, for it is a bijection and, on the one hand, $(f+g)(1)=f(1)+g(1)$ gives

$$\vartheta(f+g)=\vartheta(f)+\vartheta(g)$$

whereas, on the other hand, if $f(1)=p$ and $g(1)=q$ then $(f\circ g)(1)=pq=(g\circ f)(1)$ and consequently

$$\vartheta(f\circ g)=\vartheta(f)\vartheta(g).$$

Exactly the same argument holds in the case where $A=\mathbb{Z}_n$.

Suppose now that $A=\mathbb{Z}\times\mathbb{Z}$. If $f,g:A\to A$ are given by $f(x,y)=(x,0)$ and $g(x,y)=(x,x)$ then we have $(f\circ g)(x,y)=(x,0)$ and $(g\circ f)(x,y)=(x,x)$, so that $f\circ g\neq g\circ f$ and hence $\text{Hom}(A)$ is not commutative.

2.21 (*a*) True. If $ab=ba$ then $f(ab)=f(ba)$ so $f(a)f(b)=f(b)f(a)$. Since every element of S is of the form $f(a)$ for some $a\in R$, the result follows.

(*b*) True. The identity element 1 of R satisfies $1a=a=a1$ for all $a\in R$ so $f(1)f(a)=f(a)=f(a)f(1)$, showing that $f(1)$ is the identity of S.

(*c*) True. Use (*b*) and the uniqueness of the identity in S.

(*d*) False. For example, $\mathbb{Z}_6$ has zero divisors and there is a surjective ring morphism $f:\mathbb{Z}_6\to\mathbb{Z}_2$ given by $f(x)=x \pmod 2$.

(*e*) False. Consider the example in (*d*); $\mathbb{Z}$ is an integral domain but $\mathbb{Z}_6$ is not.

(*f*) True. If $f:R\to S$ is a ring morphism then $\text{Ker}\, f$ is an ideal of R, so if R is a field we have $\text{Ker}\, f=\{0\}$ or $\text{Ker}\, f=R$. Now we cannot have the latter since this would imply that S is the trivial ring. Thus $\text{Ker}\, f=\{0\}$ and f is an isomorphism.

2.22 It is clear that $f(x)+f(y)=f(x+y)$ for all $x,y\in\mathbb{R}$. Also,

$$\begin{bmatrix}\frac12 x & \frac12 x\\ \frac12 x & \frac12 x\end{bmatrix}\begin{bmatrix}\frac12 y & \frac12 y\\ \frac12 y & \frac12 y\end{bmatrix}=\begin{bmatrix}\frac12 xy & \frac12 xy\\ \frac12 xy & \frac12 xy\end{bmatrix}$$

shows that $f(x)f(y)=f(xy)$ for all $x,y\in\mathbb{R}$. Thus f is a ring morphism.

Now $f(x)=f(y)$ gives $x=y$ and so f is injective. Consequently we have that $F=\text{Im}\, f\cong\mathbb{R}$.

The mapping $g:\text{Mat}_{2\times 2}(\mathbb{R})\to\mathbb{R}$ described by

$$\begin{bmatrix}a & b\\ c & d\end{bmatrix}\to a+d$$

is not a ring morphism; for example, we have that

$$\begin{bmatrix}1 & 1\\ 1 & 1\end{bmatrix}\begin{bmatrix}2 & 1\\ 1 & 2\end{bmatrix}=\begin{bmatrix}3 & 3\\ 3 & 3\end{bmatrix}\to 6$$

but $\begin{bmatrix} 1 & 1 \\ 1 & 1 \end{bmatrix} \to 2$ and $\begin{bmatrix} 2 & 1 \\ 1 & 2 \end{bmatrix} \to 4.$

The mapping $g \circ f$ is the identity map on $\mathbb{R}$ and so is an isomorphism. However, $f \circ g$ is not a ring morphism, nor is it injective or surjective.

2.23 We have $R = \left\{ \frac{a}{b} \mid a, b \in \mathbb{Z}, b \notin 3\mathbb{Z} \right\}$ and the identities

$$\frac{a}{b} - \frac{c}{d} = \frac{ad - bc}{bd}, \quad \frac{a}{b} \cdot \frac{c}{d} = \frac{ac}{bd}.$$

Since 3 is prime, if 3 does not divide b and 3 does not divide d then 3 does not divide the product bd. It follows immediately that R is a subring of $\mathbb{Q}$.

If $I = \left\{ \frac{a}{b} \in R \mid 3 \text{ divides } a \right\}$ then it is clear from the above identities that I is an ideal of R.

Consider a *non-zero* element $\frac{a}{b} + I$ of the quotient ring R/I. We have $\frac{a}{b} \in R$ with $\frac{a}{b} \notin I$. It follows that $\frac{b}{a} \in R$ and $\frac{b}{a} \notin I$. Now

$$\left(\frac{a}{b} + I\right)\left(\frac{b}{a} + I\right) = \frac{ab}{ba} + I = 1 + I$$

and so $\frac{a}{b} + I$ is invertible in R/I. Hence R/I is a field.

2.24 Since $x = x^2$ for every $x \in R$ we have, for every $a \in R$,

$$a + a = (a + a)(a + a) = a^2 + a^2 + a^2 + a^2 = a + a + a + a.$$

Consequently we have that $2a = 0$. Also, for all $x, y \in R$, we have

$$x + y = (x + y)(x + y) = x^2 + xy + yx + y^2 = x + xy + yx + y,$$

which gives $xy = -yx$. But $2yx = 0$ so $yx = -yx$ and we deduce that $xy = yx$, whence R is commutative.

That Δ is associative is quickly seen by examining the Venn diagrams for $A\Delta(B\Delta C)$ and $(A\Delta B)\Delta C$; they are each as shown in Fig. S2.1.

It is now clear that $\mathbf{P}(S)$ is an abelian group under Δ, the identity element being $\emptyset$ and A being its own inverse. To check the distributive law $C \cap (A\Delta B) = (C \cap A)\Delta(C \cap B)$, consider again the Venn diagram for each side; in each case it is as shown in Fig. S2.2.

Fig.S2.1

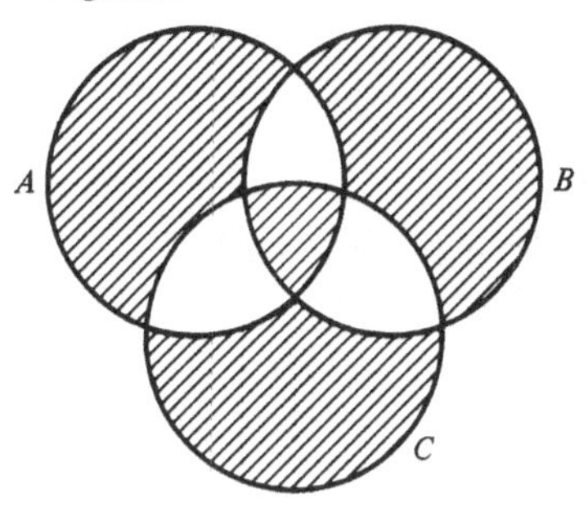

Fig.S2.2

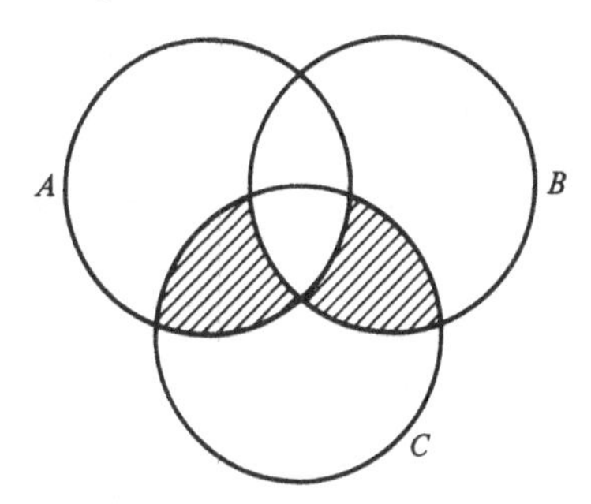

It is now clear that $\mathbf{P}(S)$ is a ring, which is boolean since $A \cap A = A$ for every $A \in \mathbf{P}(S)$.

Consider now the mapping $f_E : \mathbf{P}(S) \to \mathbf{P}(S)$ given by $f_E(X) = X \cap E'$. We have

$$\begin{aligned} f_E(X \Delta Y) = (X \Delta Y) \cap E' &= (X \cap E') \Delta (Y \cap E') \\ &= f_E(X) \Delta f_E(Y); \\ f_E(X \cap Y) = X \cap Y \cap E' &= (X \cap E') \cap (Y \cap E') \\ &= f_E(X) \cap f_E(Y). \end{aligned}$$

Thus f_E is a ring morphism. Clearly, $\operatorname{Im} f_E = \mathbf{P}(E')$, and

$$\operatorname{Ker} f_E = \{X \in \mathbf{P}(S) \mid X \cap E' = \emptyset\} = \mathbf{P}(E).$$

Hence $\mathbf{P}(E)$ is an ideal of $\mathbf{P}(S)$ and, by the first isomorphism theorem for rings,

$$\mathbf{P}(S)/\mathbf{P}(E) \cong \mathbf{P}(E').$$

2.25 We have that

$$\alpha(f)\alpha(g) = \begin{bmatrix} f(0) & 0 \\ Df(0) & f(0) \end{bmatrix} \begin{bmatrix} g(0) & 0 \\ Dg(0) & g(0) \end{bmatrix}$$
$$= \begin{bmatrix} f(0)g(0) & 0 \\ Df(0)g(0) + f(0)Dg(0) & f(0)g(0) \end{bmatrix}$$
$$= \begin{bmatrix} (fg)(0) & 0 \\ D(fg)(0) & (fg)(0) \end{bmatrix}$$
$$= \alpha(fg),$$

and clearly $\alpha(f) + \alpha(g) = \alpha(f+g)$. Hence α is a ring morphism.

Im α consists of the matrices of the form $\begin{bmatrix} a & 0 \\ b & a \end{bmatrix}$ where $a, b \in \mathbb{R}$; and

$$\operatorname{Ker} f = \{f \in \mathbb{R}[X] \mid 0 = f(0) = Df(0)\} = (X^2).$$

2.26 That R and S are subrings is routine. Now R contains an element α with $\alpha^2 = 2$; for $(0 + 1\sqrt{2})^2 = 2$. Thus if $f : R \to S$ were an isomorphism we would have

$$f(\alpha^2) = f(2) = f(1+1) = f(1) + f(1) = 1 + 1 = 2$$

whence $[f(\alpha)]^2 = 2$.

But S contains no element β with $\beta^2 = 2$. For, if $(a + b\sqrt{3})^2 = 2$ then $a^2 + 3b^2 + 2ab\sqrt{3} = 2$ and consequently $ab = 0$, so that either $a = 0$ or $b = 0$. If $a = 0$ then $3b^2 = 2$, which is impossible since $b \in \mathbb{Z}$; and if $b = 0$ then $a^2 = 2$, which is also impossible since $a \in \mathbb{Z}$.

We conclude that no such isomorphism exists.

2.27 We have that $(2n+1)^2 = 4n^2 + 4n + 1$ and so $\vartheta[(2n+1)^2] = 1$.

Every member of the given sequence is of the form $100k + 11$ for some $k \in \mathbb{Z}$. But

$$\vartheta(100k + 11) = \vartheta(100)\vartheta(k) + \vartheta(11) = 0.\vartheta(k) + 3 = 3.$$

Hence no member of the sequence is an odd square and therefore no member can be a square.

2.28 The only two rings with p elements (where p is prime) must each have additive group $\mathbb{Z}_p$ and either zero multiplication or the usual multiplication in $\mathbb{Z}_p$. To prove that these are the only possibilities, let g be an additive generator of $\mathbb{Z}_p$ and suppose that the multiplication is not the zero multiplication, so that $g^2 = kg$ for some integer $k \neq 0$. Then we can find an integer m such that

$km \equiv 1 \pmod{p}$. Consider now the element $h = mg = g + g + \cdots + g$ (m summands). We have $h^2 = m^2g^2 = m^2kg = mg = h$. If we call the ring R_p then we can consider the mapping f from the ring $\mathbb{Z}_p$ to the ring R_p described by $f(n) = nh$. We have

$$f(n+m) = (n+m)h = nh + mh;$$
$$f(nm) = (nm)h = nmh^2 = nhmh$$

and so f is a ring morphism. Since every non-zero element of the group $\mathbb{Z}_p$ is an additive generator of $\mathbb{Z}_p$, this is also true of h. Consequently, f is surjective. Since the sets in question are finite and have the same number of elements, f is also injective. Hence f is an isomorphism and the multiplication in R_p is the usual multiplication.

2.29 (*a*) True. If R is a subring of a field F and $a, b \in R$ with $ab = 0$ then $a, b \in F$ with $ab = 0$, so $a = 0$ or $b = 0$.

(*b*) False. For example, $\mathbb{Z}$ is an integral domain but the quotient ring $\mathbb{Z}/4\mathbb{Z} \cong \mathbb{Z}_4$ is not an integral domain.

(*c*) False. For example, $\mathbb{Q}$ is an integral domain in which every non-zero element is invertible. But $\mathbb{Q} \setminus \{0\}$ cannot be cyclic, since if it were generated by $r \in \mathbb{Q} \setminus \{0\}$ we would have $r^n = 1$ for some n, which would imply that $r = \pm 1$, which is impossible.

(*d*) True. If R satisfies the given property then $R \setminus \{0\}$ is generated by a as a multiplicative group. Thus, given $x, y \in R \setminus \{0\}$, we have $x = a^n$, $y = a^m$ for some n, m and so $xy = yx$. Since 0 trivially commutes with every element of R, it follows that R is commutative. Now R has an identity, so $a^n = 1$ for some n since $R \setminus \{0\}$ is cyclic. Therefore every non-zero element of R is invertible and so R is a field.

(*e*) True. Take, for example, $R = F$.

2.30 (*a*) If $a^m = 0$ and $b^n = 0$, consider $(a-b)^{m+n}$. Since R is commutative, this can be expanded by the binomial theorem. Each term of the expansion contains a power of a no less than m or a power of b no less than n. Hence each term is zero and so $(a-b)^{n+m} = 0$. Thus we see that if N denotes the set of nilpotent elements then $a, b \in N$ implies $a - b \in N$. Also, if $a \in N$ then $a^m = 0$ implies, since R is commutative, that $(ra)^m = 0$ for every $r \in R$. Hence $ra \in N$ and so N is an ideal.

(*b*) Suppose that $r + N$ is nilpotent in R/N. Then we have, for some m, $(r+N)^m = N$. It follows that $r^m + N = N$ and so $r^m \in N$, whence

$r^{mn} = (r^m)^n = 0$ for some n, and consequently $r \in N$. Thus $r + N$ is the zero of R/N.

(*c*) First try some examples. If $n = 6$ then $N = \{0\}$; if $n = 4$ then $N = \{0, 2\}$; if $n = 12$ then $N = \{0, 6\}$. In general, if

$$n = p_1^{\alpha_1} \dots p_k^{\alpha_k}$$

with each p_i a prime and each $\alpha_i \geqslant 1$, then

$$N = \{0\} \cup \{p_1^{\beta_1} \dots p_k^{\beta_k} \mid 1 \leqslant \beta_i \leqslant \alpha_i\}.$$

(*d*) The matrices $\begin{bmatrix} 0 & 1 \\ 0 & 0 \end{bmatrix}$ and $\begin{bmatrix} 0 & 0 \\ 1 & 0 \end{bmatrix}$ are both nilpotent (each satisfies $x^2 = 0$), but their sum is not nilpotent. Thus N is not necessarily an ideal when R is non-commutative.

2.31 Consider the rectangular array (with $m + 1$ rows and $n + 1$ columns) whose (i, j)th position is occupied by $a_i b_j x^{i+j} = a_i x^i . b_j x^j$. Then

$$\sum_{i=0}^{n} \sum_{j=0}^{m} a_i b_j x^{i+j}$$

is the sum of all the elements in this array. Now extend the array to an array of size $(m + n + 1) \times (m + n + 1)$ by defining $a_i = b_j = 0$ for $i \geqslant n + 1$ and $j \geqslant m + 1$. Then

$$\sum_{k=0}^{n+m} \left(\sum_{j=0}^{k} a_{k-j} b_j \right) x^k$$

is the sum of the elements in the second array, this being achieved by summing its SW-NE diagonals. Clearly, because of the zero entries in the second array, the sums are the same.

To show that ζ_x is a ring morphism it is clearly enough to verify that it preserves products. Now if

$$f(X) = a_0 + a_1 X + \dots + a_n X^n,$$
$$g(X) = b_0 + b_1 X + \dots + b_m X^m$$

then the product $f(X)g(X)$ is, by definition, the polynomial

$$\sum_{k=0}^{n+m} \left(\sum_{j=0}^{k} a_{k-j} b_j \right) X^k.$$

The result is then immediate from the previous identity.

ζ_x is surjective since for every $a \in \mathbb{R}$ we have

$$a = \zeta_x(a + 0X + \dots + 0X^n).$$

Now $f(X) \in \operatorname{Ker} \zeta_x$ if and only if $X - x$ divides $f(X)$. Thus Ker ζ_x is the ideal generated by $X - x$ (i.e. it is the smallest ideal containing the polynomial $X - x$).

2.32 It is clear that $\bar{\vartheta}_m$ preserves sums. As for products, we have, for $f(X) = a_0 + a_1X + \cdots + a_nX^n$ and $g(X) = b_0 + b_1X + \cdots + b_pX^p$,

$$\begin{aligned}
\bar{\vartheta}_m(f(X)g(X)) &= \bar{\vartheta}_m\left[\sum_{k=0}^{n+p}\left(\sum_{j=0}^{k} a_{k-j}b_j\right)X^k\right] \\
&= \sum_{k=0}^{n+p}\left(\sum_{j=0}^{k} \vartheta_m(a_{k-j}b_j)\right)X^k \\
&= \sum_{k=0}^{n+p}\left(\sum_{j=0}^{k} \vartheta_m(a_{k-j})\vartheta_m(b_j)\right)X^k \\
&= \left(\sum_{j=0}^{n} \vartheta_m(a_i)X^i\right)\left(\sum_{j=0}^{p} \vartheta_m(b_j)X^j\right) \\
&= \bar{\vartheta}_m[f(X)]\bar{\vartheta}_m[g(X)].
\end{aligned}$$

Thus $\bar{\vartheta}_m$ is a ring morphism. Since ϑ_m is surjective, so also is $\bar{\vartheta}_m$. The kernel of $\bar{\vartheta}_m$ consists of all polynomials $\Sigma_{i=0}^{n} a_iX^i$ such that $\vartheta_m(a_i) = 0$ in $\mathbb{Z}_m$ for every i; i.e. of all polynomials of the form $m(\Sigma_{i=0}^{n} b_iX^i)$.

(*a*) Taking coefficients in $\mathbb{Z}_5$ we obtain, by long division,

$$X^4 + 3X^3 + 2X^2 + X + 4 = (3X^2 + 2X)(2X^2 + 3X + 2) \\ -3X + 4$$

so the quotient is $2X^2 + 3X + 2$ and the remainder is $-3X + 4 = 2X + 4$.

(*b*) In $\mathbb{Z}[X]$ we have

$$X^{10} + 1 = (X^2 + 1)(X^8 - X^6 + X^4 - X^2 + 1).$$

Applying the morphism $\bar{\vartheta}_2$ we obtain, in $\mathbb{Z}_2[X]$,

$$X^{10} + 1 = (X^2 + 1)(X^8 + X^6 + X^4 + X^2 + 1).$$

Hence the quotient is $X^8 + X^6 + X^4 + X^2 + 1$ and the remainder is $-1 = 1$.

For the last part of the question we note that in $\mathbb{Z}[X]$ we have

$$X^5 - 10X + 12 = (X^3 - 2X)(X^2 + 2) - 6X + 12.$$

Applying the morphism $\bar{\vartheta}_n$ to this, we see that $X^2 + 2$ divides $X^5 - 10X + 12$ in $\mathbb{Z}_n[X]$ if and only if $-6X + 12 = 0$ in $\mathbb{Z}_n[X]$. This is the case if and only if $6 \equiv 0 \equiv 12 (\operatorname{mod} n)$; i.e. if and only if $n = 2, 3$ or 6.

2.33 (*a*) $\mathbb{Z}_6[X]$ has zero divisors; for example, $2X.3X = 0$ but $2X \neq 0$ and $3X \neq 0$. If $f(X) = 1 + a_1X + \cdots + a_nX^n$ then $f(X)$ cannot be a zero divisor; for, if $g(X) = b_1X + \cdots + b_mX^m$ has constant term zero then

$$f(X)g(X) = b_1X + \cdots + a_nb_mX^{n+n} \neq 0.$$

If $f(X)$ has constant term a where a is not a zero divisor in R then $f(X)$ is not a zero divisor in $R[X]$.

(*b*) If $1+aX$ is invertible then there exists $b_0 + b_1X + \cdots + b_mX^m \in R[X]$ with

$$(1 + aX)(b_0 + b_1X + \cdots + b_mX^m) = 1.$$

Thus we have that

$$b_0 + (b_1 + ab_0)X + (b_2 + ab_1)X^2 + \cdots + (b_m + ab_{m-1})X^m + ab_mX^{m+1} = 1$$

from which we obtain

$$b_0 = 1, \quad b_1 = -a, \quad b_2 = a^2, \quad b_3 = -a^3, \ldots, b_m = \pm a^m,$$
$$\text{and } ab_m = 0.$$

Then $ab_m = 0$ gives $a^{m+1} = 0$.

Conversely, if $a^n = 0$ for some n then it is easy to check that

$$1 - aX + a^2X^2 - a^3X^3 + \cdots \pm a^{n-1}X^{n-1}$$

is the inverse of $1 + aX$.

(*c*) If F is a field then the non-zero constant polynomials are the only invertible elements of $F[X]$; for we have

$$\deg(f(X)g(X)) = \deg f(X) + \deg g(X)$$

and so $\deg(f(X)g(X)) = 0$ only if $\deg f(X) = \deg g(X) = 0$.

2.34 Let $p(X) = a_nX^n + \cdots + a_0$ and $q(X) = b_mX^m + \cdots + b_0$ with $a_n \neq 0$ and $b_m \neq 0$. Then $p(X)q(X) = a_nb_mX^{n+m} + \cdots + a_0b_0$ and, since R is an integral domain, $a_nb_m \neq 0$. Hence the product of two non-zero polynomials is non-zero and so $R[X]$ is an integral domain.

Consider now the polynomial X. If X had a multiplicative inverse we would have

$$a_nX^{n+1} + \cdots + a_0X = X(a_nX^n + \cdots + a_0) = 1$$

which is impossible. Hence $R[X]$ is never a field.

Now let $R = \mathbb{Z}_4$. Suppose that the polynomial $a_nX^n + \cdots + a_0$ has inverse $b_mX^m + \cdots + b_0$. Then we must have $a_0b_0 = 1$ and so, since 1 and 3 are the only elements in $\mathbb{Z}_4$ with multiplicative inverses, it follows that either $a_0 = 1$ or $a_0 = 3$.

If $p(X) = a^n X^n + \cdots + a_0$ is such that $a_0 = 1$ or 3 and, for $i \geqslant 1$, $a_i = 0$ or 2 then we can write $p(X)$ in the form $a_0 + 2q(X)$ for some $q(X) \in \mathbb{Z}_4[X]$. Since in $\mathbb{Z}_4$ we have $1^2 = 3^2 = 1$ and $4 = 0$, it follows that

$$[p(X)]^2 = (a_0 + 2q(X))^2 = a_0^2 + 4a_0 q(X) + 4[q(X)]^2 = 1.$$

Hence $p(X)$ is invertible.

2.35 (*a*) Let $f(X) = X^2$ and $g(X) = 2X$ in $\mathbb{Z}_4[X]$. Suppose that

$$X^2 = 2X[q(X)] + r(X).$$

Then we must have $\deg r(X) < 1$ and so $r(X)$ is a constant polynomial. But $2X^2 = 2r(X)$, which provides a contradiction.

(*b*) Let $f(X) = X^2$ and $g(X) = 2X$ in $\mathbb{Z}[X]$. Again suppose that

$$X^2 = 2X[q(X)] + r(X).$$

Then, since $\mathbb{Z}$ is an integral domain, $\deg q(X) = 1$ and so

$$X^2 = 2X(aX + b) + r$$

which gives $2a = 1$, a contradiction since the coefficients are in $\mathbb{Z}$.

2.36 Use the euclidean algorithm:

(*a*) $\text{hcf} = X - 1; a(X) = \frac{1}{7}(X^2 - 3X + 3); b(X) = \frac{1}{7}(-X + 2);$

(*b*) $\text{hcf} = X - \text{i}; a(X) = -X; b(X) = 1;$

(*c*) $\text{hcf} = X^2 + X + 2; a(X) = 0; b(X) = 1.$

2.37 First observe that if $K = \begin{bmatrix} 1 & 1 \\ -1 & 0 \end{bmatrix}$ then

$$K^2 = K - I_2.$$

Now $(aI_2 + bK) - (cI_2 + dK) = (a - c)I_2 + (b - d)K$ and so F is stable under subtraction. Also, using the above observation, we have

$$\begin{aligned}(aI_2 + bK)(cI_2 + dK) &= acI_2 + adK + bcK + bdK^2 \\ &= acI_2 + adK + bcK + bdK - bdI_2 \\ &= (ac - bd)I_2 + (ad + bc + bd)K\end{aligned}$$

so that F is also stable under multiplication. Hence F is a subring of $\text{Mat}_{2\times 2}(\mathbb{R})$. Swapping a, c and b, d in the above multiplication, we see that F is commutative. Now

$$aI_2 + bK = \begin{bmatrix} a & 0 \\ 0 & a \end{bmatrix} + \begin{bmatrix} b & b \\ -b & 0 \end{bmatrix} = \begin{bmatrix} a+b & b \\ -b & a \end{bmatrix}$$

and the determinant of this is $a^2 + ab + b^2$. We now observe that

$$a^2 + ab + b^2 = 0 \Leftrightarrow a = 0 = b.$$

(In fact, if $a^2+ab+b^2=0$ then we have on the one hand $ab=-a^2-b^2\leqslant 0$, and on the other hand $ab=(a+b)^2\geqslant 0$ so that $ab=0$ and hence $a=0=b$.) From this observation it follows that aI_2+bK has zero determinant (i.e. is not invertible) if and only if $a=0=b$. Thus we see that every non-zero element of F has an inverse in F, so F is a subfield of $\mathrm{Mat}_{2\times 2}(\mathbb{R})$.

There is no $q\in\mathbb{Q}$ such that $q^2-q+1=0$ since the roots of $x^2-x+1=0$ are complex. Thus X^2-X+1 is irreducible over $\mathbb{Q}$.

Now $\mathbb{Q}[X]/(X^2-X+1)$ is a field which can be regarded as $\{a+bx \mid a,b\in\mathbb{Q},\ x^2=x-1\}$. This is clearly isomorphic to F under the assignment $a+bx\leftrightarrow aI_2+bK$, since $K^2=K-I_2$.

2.38 That R forms a subring of $\mathrm{Mat}_{3\times 3}(\mathbb{R})$ is routine. That α is a ring morphism is also straightforward. Clearly, α is surjective. By the first isomorphism theorem we have that $R/\mathrm{Ker}\,\alpha\cong\mathrm{Im}\,\alpha$. The required ideal is therefore $\mathrm{Ker}\,\alpha$, which consists of the matrices of the form

$$\begin{bmatrix}0 & a & b\\ 0 & 0 & c\\ 0 & 0 & 0\end{bmatrix}$$

where $a,b,c\in\mathbb{R}$.

2.39 The product is

$$\begin{bmatrix}6a+3b-2c-d & 3a+3b-c-d & 3a+6b-c-2d\\ -10a-5b+2c+d & -5a-5b+c+d & -5a-10b+c+2d\\ 2a+b+2c+d & a+b+c+d & a+2b+c+2d\end{bmatrix}.$$

It follows that

$$\vartheta\left(\begin{bmatrix}a & b\\ c & d\end{bmatrix}+\begin{bmatrix}p & q\\ r & s\end{bmatrix}\right)$$

$$=\tfrac{1}{2}\begin{bmatrix}3 & -1\\ -5 & 1\\ 1 & 1\end{bmatrix}\left(\begin{bmatrix}a & b\\ c & d\end{bmatrix}+\begin{bmatrix}p & q\\ r & s\end{bmatrix}\right)\begin{bmatrix}2 & 1 & 1\\ 1 & 1 & 2\end{bmatrix}$$

$$=\vartheta\begin{bmatrix}a & b\\ c & d\end{bmatrix}+\vartheta\begin{bmatrix}p & q\\ r & s\end{bmatrix}.$$

Also, since

$$\begin{bmatrix} 2 & 1 & 1 \\ 1 & 1 & 2 \end{bmatrix} \begin{bmatrix} 3 & -1 \\ -5 & 1 \\ 1 & 1 \end{bmatrix} = \begin{bmatrix} 2 & 0 \\ 0 & 2 \end{bmatrix}$$

it is similarly seen that ϑ preserves multiplication. Hence ϑ is a ring morphism.

Now $\begin{bmatrix} a & b \\ c & d \end{bmatrix} \in \operatorname{Ker} \vartheta$ gives

$$\begin{bmatrix} 3 & -1 \\ -5 & 1 \\ 1 & 1 \end{bmatrix} \begin{bmatrix} a & b \\ c & d \end{bmatrix} \begin{bmatrix} 2 & 1 & 1 \\ 1 & 1 & 2 \end{bmatrix} = 0.$$

Multiplying on the left by $\begin{bmatrix} 2 & 1 & 1 \\ 1 & 1 & 2 \end{bmatrix}$ and on the right by $\begin{bmatrix} 3 & -1 \\ -5 & 1 \\ 1 & 1 \end{bmatrix}$ produces $\begin{bmatrix} a & b \\ c & d \end{bmatrix} = 0$, from which we deduce that $\operatorname{Ker} \vartheta = \{0\}$. By the first isomorphism theorem we then have that $\operatorname{Im} \vartheta \simeq \operatorname{Mat}_{2\times 2}(\mathbb{R})$.

2.40 We have that

$$X^5 + X^4 + 2X^3 + 2X^2 + 2X + 1 = (X+1)(X^4 + X^2 + 1) + X^3 + X^2 + X$$

$$X^4 + X^2 + 1 = (X-1)(X^3 + X^2 + X) + X^2 + X + 1$$

$$X^3 + X^2 + X = X(X^2 + X + 1).$$

Hence $h = X^2 + X + 1$.

Clearly $f, g \in \{\alpha h \mid \alpha \in \mathbb{Q}[X]\}$. But any ideal containing f and g contains h since

$$X^2 + X + 1 = X^2(X^4 + X^2 + 1) + (1 - X)(X^5 + X^4 + 2X^3 + 2X^2 + 2X + 1).$$

Consider the mapping $\zeta : \mathbb{Q}[X] \to \mathbb{C}$ induced by substituting $e^{2\pi i/3}$ for X. Clearly, $\operatorname{Ker} \zeta = I$. Also

$$\begin{aligned} \operatorname{Im} \zeta &= \left\{ p + q\left(\cos \frac{2\pi}{3} + i \sin \frac{2\pi}{3}\right) \mid p, q \in \mathbb{Q} \right\} \\ &= \left\{ p - \tfrac{1}{2}q + q \frac{\sqrt{3}}{2} i \mid p, q \in \mathbb{Q} \right\} \\ &= \{ a + b\sqrt{3}\, i \mid a, b \in \mathbb{Q} \}. \end{aligned}$$

The result now follows by the first isomorphism theorem.

2.41 Working in $\mathbb{Z}_5[X]$ we have

$$\begin{array}{llll}
 & f = X^4 + X^3 + 4X^2 + X - 2 & X^3 + X^2 \quad -2 = g & \\
 & Xg = X^4 + X^3 \quad -2X & & \\
\hline
f - Xg = & 4X^2 + 3X - 2 \longrightarrow & 4X^3 + 3X^2 - 2X & = X(f - Xg) \\
\hline
 & 4X^2 - 2X - 2 \longleftarrow & 4X^2 - 2X - 2 & = g + X(f - Xg) \\
\hline
 & 0 & &
\end{array}$$

The highest common factor of $f(X)$, $g(X)$ in $\mathbb{Z}_5[X]$ is thus $4^{-1}(4X^2 - 2X - 2)$. Now in $\mathbb{Z}_5$ we have $4^{-1} = 4$, so the highest common factor is

$$16X^2 - 8X - 8 = X^2 - 3X - 3 = X^2 + 2X + 2.$$

Working in $\mathbb{Z}_7[X]$ we have

$$\begin{array}{llll}
 & f = X^4 + X^3 + 4X^2 + 4X - 2 & X^3 + X^2 + 5X \quad -2 = g & \\
 & Xg = X^4 + X^3 + 5X^2 - 2X & & \\
\hline
f - Xg = & -X^2 + 6X - 2 \longrightarrow & -X^3 + 6X^2 - 2X & = X(f - Xg) \\
\hline
 & 6X^2 - 4X \longleftarrow & 3X - 2 & = g + X(f - Xg) \\
\hline
 & 3X - 2 \longrightarrow & 3X - 2 & \\
\hline
 & & 0 &
\end{array}$$

The highest common factor of $f(X)$, $g(X)$ in $\mathbb{Z}_7[X]$ is thus $3^{-1}(3X - 2)$. Now in $\mathbb{Z}_7$ we have $3^{-1} = 5$, so the highest common factor is

$$5(3X - 2) = 15X - 10 = X - 3 = X + 4.$$

2.42 For the matrix $K = \begin{bmatrix} 0 & q \\ 1 & 0 \end{bmatrix}$ we have that $K^2 = qI_2$. Thus

$$\begin{aligned}
(aI_2 + bK)(cI_2 + dK) &= acI_2 + adK + bcK + bdK^2 \\
&= acI_2 + adK + bcK + bdqI_2 \\
&= (ac + bdq)I_2 + (ad + bc)K.
\end{aligned}$$

This shows that S_q is stable under multiplication. It is also stable under subtraction since

$$aI_2 + bK - (cI_2 + dK) = (a - c)I_2 + (b - d)K.$$

Thus we see that S_q is a subring of $\mathrm{Mat}_{2\times 2}(\mathbb{Q})$. Clearly, I_2 is the identity element of S_q; and from the above multiplication we see that S_q is commutative.

Now

$$aI_2 + bK = \begin{bmatrix} a & 0 \\ 0 & a \end{bmatrix} + b\begin{bmatrix} 0 & q \\ 1 & 0 \end{bmatrix} = \begin{bmatrix} a & bq \\ b & a \end{bmatrix}$$

and so $aI_2 + bK = 0$ if and only if $a = b = 0$. Also,

$$\det(aI_2 + bK) = a^2 - b^2 q$$

and this is zero if and only if $a = b\sqrt{q}$. Thus we see that every non-zero element of S_q has an inverse (i.e. S_q is a field) if and only if $\sqrt{q} \notin \mathbb{Q}$.

If $\sqrt{q} \notin \mathbb{Q}$ then clearly $X^2 - q$ is irreducible over $\mathbb{Q}$ and consequently $\mathbb{Q}[X]/(X^2 - q)$ is a field. This field can be regarded as

$$\{a + bx \mid a, b \in \mathbb{Q}, x^2 = q\}.$$

Clearly, since $K^2 = qI_2$, this is isomorphic to the field S_q under the assignment $a + bx \leftrightarrow aI_2 + bK$.

2.43 Suppose that $x \in R$ is such that x^{-1} exists. If $x \in I$ then we have $1 = x^{-1}x \in I$ and then $(\forall r \in R) r = r1 \in I$ so that $R = I$.

It is clear that each of I_1, I_2, I_3 is stable under subtraction and multiplication, so that each is a subring of R.

Consider now the general products

$$\begin{bmatrix} a & b \\ 0 & c \end{bmatrix}\begin{bmatrix} x & y \\ 0 & z \end{bmatrix} = \begin{bmatrix} ax & ay + bz \\ 0 & cz \end{bmatrix},$$

$$\begin{bmatrix} x & y \\ 0 & z \end{bmatrix}\begin{bmatrix} a & b \\ 0 & c \end{bmatrix} = \begin{bmatrix} xa & xb + yc \\ 0 & zc \end{bmatrix}.$$

Taking $z = 0$ we see that the products are in I_1 and so I_1 is an ideal of R.
Taking $x = 0$ we see that the products are in I_2 and so I_2 is an ideal of R.
Taking $x = z = 0$ we see that the products are in I_3 and so I_3 is an ideal of R.

Suppose now that I is an ideal of R with $I \neq \{0\}$. If I contains a matrix of the form $\begin{bmatrix} x & y \\ 0 & z \end{bmatrix}$ with $x \neq 0$ and $z \neq 0$ then, since this matrix is invertible in R, it follows by the first part of the question that $I = R$.

Suppose now that $I \neq R$ and that $I \neq \{0\}$. If I contains $\begin{bmatrix} x & y \\ 0 & 0 \end{bmatrix}$ with $x \neq 0$ then I contains all left/right multiples of this by elements of R, whence I contains I_1. If $I_1 \subset I$ then there exists $\begin{bmatrix} x & y \\ 0 & z \end{bmatrix}$ in I with $x, z \neq 0$ whence the contradiction $I = R$. Hence in this case we have $I = I_1$. A similar argument shows that if I contains $\begin{bmatrix} 0 & y \\ 0 & z \end{bmatrix}$ with $z \neq 0$ then I contains I_2 and in fact coincides with I_2. Clearly, the only other possibility for I (other than $\{0\}$, R, I_1, I_2) is $I = I_3$.

2.44 Since we are dealing with $\mathbb{Z}_2$, all coefficients are either 0 or 1. If $f(X)$ has an even number of non-zero coefficients then 1 is a root of $f(X) = 0$, so $X + 1$ is a factor of $f(X)$. Thus, if $f(X)$ is irreducible then it must have an odd number of non-zero coefficients. Clearly X is irreducible of degree 1. Also, any polynomial of degree greater than or equal to 2 with zero constant term is not irreducible. Consider now polynomials with constant term 1.

There is only one irreducible polynomial in $\mathbb{Z}_2[X]$ of degree 1, namely $X+1$. Hence the only polynomial of degree 2 that is not irreducible is $(X+1)(X+1) = X^2+1$. Hence X^2+X+1 is the only irreducible of degree 2. Similarly, $(X^2+X+1)(X+1)$ and $(X+1)^3$ are the only polynomials of degree 3 that fail to be irreducible, so X^3+X+1 and X^3+X^2+1 are the irreducibles of degree 3. Likewise,

$$X^4+X+1, \quad X^4+X^3+1, \quad X^4+X^3+X^2+X+1$$

are the irreducibles of degree 4.

Examples of fields with 4,8,16 elements are $\mathbb{Z}_2[X]/(p(X))$ where $p(X)$ is $X^2+X+1, X^3+X+1, X^4+X+1$ respectively.

The six given polynomials all have degree less than or equal to 8, so are either irreducible or have a factor which is irreducible and of degree less than or equal to 4. We obtain

(*a*) $X^5+X^4+1 = (X^2+X+1)(X^3+X+1)$;

(*b*) $X^6+X^5+X^2+X+1$ is irreducible;

(*c*) $X^6+X^4+X^3+X^2+1 = (X^2+X+1)(X^4+X^3+X^2+X+1)$;

(*d*) $X^7+X^6+X^4+X^3+1 = (X^3+X+1)(X^4+X^3+X^2+X+1)$;

(*e*) $X^7+X^6+X^3+X+1$ is irreducible;

(*f*) $X^8+X^6+X^5+X^4+X^3+X^2+1 = (X^2+X+1)(X^3+X+1)(X^3+X^2+1)$.

The polynomial X^4+X+1 is irreducible in $\mathbb{Z}_2[X]$ so the quotient $\mathbb{Z}_2[X]/(X^4+X+1)$ is a field with 16 elements; for it can be regarded as

$$\{a+bx+cx^2+dx^3 \mid a,b,c,d \in \mathbb{Z}_2, x^4 = x+1\}.$$

The powers of X in the image are

$$X, X^2, X^3, X+1, X^2+X, X^3+X^2, X^3+X+1,$$
$$X^2+1, X^3+X, X^2+X+1, X^3+X^2+X,$$
$$X^3+X^2+X+1, X^3+X^2+1, X^3+1, 1.$$

These elements exhaust the non-zero elements of the field.

However, in the quotient $\mathbb{Z}_2[X]/(X^4 + X^3 + X^2 + X + 1)$ the powers of X are

$$X, X^2, X^3, X^3 + X^2 + X + 1, 1$$

which generate a proper subgroup of the multiplicative group. This multiplicative group is cyclic, however, since $X + 1$ is a generator.

Test paper 1

Time allowed: 3 hours
(Allocate 20 marks for each question.)

1 If G is a group then $x \in G$ is said to be a commutator, if for some $a, b \in G$,

$$x = [a, b] = a^{-1}b^{-1}ab.$$

Determine the set H of commutators in S_3 and show that it is a normal subgroup of S_3.

Prove that K is a normal subgroup of a group G if and only if $[k, g] \in K$ for every $k \in K$ and every $g \in G$.

2 Let G be the multiplicative group of non-singular $n \times n$ matrices with rational entries. Let $f : S_n \to G$ be defined as follows : for every permutation π let

$$f(\pi) = [a_{ij}] \text{ where } a_{ij} = \begin{cases} 1 \text{ if } \pi(j) = i; \\ 0 \text{ if } \pi(j) \neq i. \end{cases}$$

Prove that f is a group morphism and that $\text{Ker} f = \{1\}$. If $\pi = (k\ l)$ is a transposition prove that $\det f(\pi) = -1$. Hence show that no permutation $\pi \in S_n$ which can be written as a product of an even number of transpositions may be written as a product of an odd number of transpositions.

3 Show that

$$H = \{(1), (12)(34), (13)(24), (14)(23)\}$$

is a normal subgroup of S_4. Let K consist of those elements of S_4 which leave 4 fixed. Prove that $K \cong S_3$ and that $H \cap K = \{1\}$.

Show that if $h_1, h_2 \in H$ and $k_1, k_2 \in K$ then $h_1k_1 = h_2k_2$ implies $h_1 = h_2$ and $k_1 = k_2$. Deduce that any element $\pi \in S_4$ can be written uniquely as

$\pi = hk$ where $h \in H$ and $k \in K$. Use this fact to construct a surjective morphism from S_4 to K, and deduce that S_4 has both a subgroup and a quotient group isomorphic to S_3.

4 Let p be a fixed prime and let R be the set of all rationals expressible in the form m/p^k where $m, k \in \mathbb{Z}$. Prove that R is a subring of the ring $\mathbb{Q}$. Is R an integral domain? Does R contain a 1?

Find all the invertible elements of R and show that R is not a field.

For every $a \in R$ define $aR = \{ax \mid x \in R\}$. Prove that aR is a two-sided ideal of R. Suppose now that q is a prime with $q \neq p$. Show that qR is a proper subset of R whereas pR coincides with R.

Show finally that if p, q_1, q_2 are distinct primes then

$$q_1 R \cap q_2 R = (q_1 q_2)R.$$

5 Let $A \in \mathrm{Mat}_{n \times n}(\mathbb{Q})$ and let E_{ij} be the matrix whose only non-zero element is a 1 in the (i,j)th position. Calculate $E_{lm} A E_{rs}$.

Let I be a non-zero ideal of the ring $\mathrm{Mat}_{n \times n}(\mathbb{Q})$. Deduce from the above that $I = \mathrm{Mat}_{n \times n}(\mathbb{Q})$.

Show that $\vartheta : \mathrm{Mat}_{2 \times 2}(\mathbb{Q}) \to \mathrm{Mat}_{3 \times 3}(\mathbb{Q})$ defined by

$$\vartheta\left(\begin{bmatrix} a & b \\ c & d \end{bmatrix}\right) = \begin{bmatrix} -c & c & c+d \\ -a & a & a+b \\ -c & c & c+d \end{bmatrix}$$

is a ring morphism. Deduce that

$$S = \left\{ \begin{bmatrix} -c & c & c+d \\ -a & a & a+b \\ -c & c & c+d \end{bmatrix} \mid a, b, c, d \in \mathbb{Q} \right\}$$

is a subring of $\mathrm{Mat}_{3 \times 3}(\mathbb{Q})$ that is isomorphic to $\mathrm{Mat}_{2 \times 2}(\mathbb{Q})$. What is the identity element of S?

Test paper 2

Time allowed: 3 hours
(Allocate 20 marks for each question.)

1 Let $\mathbb{Q}$ be the additive group of rationals. Consider the group $G = \mathbb{Q}/\mathbb{Z}$. Show that G is infinite. Show also that every element of G has finite order.

Show that, for every integer $n \geqslant 2$, G has an element of order n. Show also that any finite collection of elements of order n lie in a cyclic subgroup of order n.

2 Let G be the group of 2×2 matrices of determinant 1 with entries in $\mathbb{Z}_3$. Show that if

$$x = \begin{bmatrix} 1 & 1 \\ 1 & 2 \end{bmatrix}, \quad y = \begin{bmatrix} 0 & 1 \\ 2 & 0 \end{bmatrix}$$

then $x, y \in G$. Prove that $x^4 = 1$, $x^2 = y^2$ and $xyx = y$. Deduce that the smallest subgroup H of G containing x and y has order 8. Prove that H is not abelian but that every subgroup of H is normal in H.

3 Let G be a finite group with an isomorphism $\vartheta : G \to G$ satisfying $\vartheta^2 = \mathrm{id}_G$ and the property that if $g \neq 1$ then $\vartheta(g) \neq g$. Show, by multiplying on the left by $[\vartheta(b)]^{-1}$ and on the right by a, that

$$\vartheta(a)a^{-1} = \vartheta(b)b^{-1} \quad \text{implies} \quad a = b.$$

Deduce that if $g \in G$ then $g = \vartheta(a)a^{-1}$ for some $a \in G$ and show that $\vartheta(g) = g^{-1}$. Prove that G is an abelian group of odd order.

4 Let m_1, m_2 be fixed positive integers which are coprime. Show that for any given integers a_1, a_2 there exists an integer y such that

$$a_1 + ym_1 \equiv a_2 \pmod{m_2}$$

and deduce that the simultaneous congruences

$$x \equiv a_1 \pmod{m_1}, \quad x \equiv a_2 \pmod{m_2}$$

have a solution. Show moreover that this solution is unique modulo $m_1 m_2$.

Let $R = \mathbb{Z}_{m_1} \times \mathbb{Z}_{m_2}$ with the ring operations

$$(a, b) + (c, d) = (a + c, b + d),$$
$$(a, b)(c, d) = (ac, bd),$$

where a, $c \in \mathbb{Z}_{m_1}$ and b, $d \in \mathbb{Z}_{m_2}$. Prove that R is isomorphic to the ring $\mathbb{Z}_{m_1 m_2}$.

5 Show that if I and J are ideals of a ring R then $I + J = \{i + j \mid i \in I, j \in J\}$ and $I \cap J$ are ideals of R.

For every $\alpha \in \mathbb{Q}$ let $I_\alpha = \{f \in \mathbb{Q}[X] \mid f(\alpha) = 0\}$. Show that I_α is an ideal of $\mathbb{Q}[X]$. Show also that if $\alpha \neq \beta \in \mathbb{Q}$ then every element of $\mathbb{Q}[X]$ can be expressed in the form $(X - \alpha)h(X) + (X - \beta)g(X)$ for some $h(X)$, $g(X) \in \mathbb{Q}[X]$.

Show that the assignment

$$(X - \alpha)h(X) + (X - \beta)g(X) + I_\beta \to (X - \alpha)h(X) + (I_\alpha \cap I_\beta)$$

defines an isomorphism from the ring $\mathbb{Q}[X]/I_\beta$ to the ring $I_\alpha/(I_\alpha \cap I_\beta)$.

Test paper 3

Time allowed: 3 hours
(Allocate 20 marks for each question.)

1 Let p be a prime and let G be the set of triples (x, y, z) of integers modulo p. Define a multiplication on G by

$$(x, y, z)(x', y', z') = (x + x', y + y', xy' + z + z').$$

Prove that G is a group of order p^3. Show also that G is not abelian and that the set H of elements of the form (x, y, z) is an abelian subgroup of order p^2.

2 Find the eight elements of the group G generated by the permutations

$$a = (1234), \quad b = (24).$$

Show that $H = \{1, b\}$ is a subgroup of G, and determine the left and right cosets of H in G. Is H a normal subgroup of G? Show also that 1 and a^2 are the only elements that commute with every element of G.

3 Let A be a ring. For every non-zero $k \in A$ let

$$S_k = \{ka \mid a \in A\}.$$

Prove that S_k is a subring of A. Show also that if A has an identity element and if k has a multiplicative inverse in A then S_k coincides with A.

Determine all the subrings of the ring of integers modulo 6.

4 Let D be an algebraic system that satisfies all the axioms for a ring except perhaps for commutativity of addition. By expanding $(a + b)(c + d)$ in two ways using the distributive laws, show that

$$ad + bc = bc + ad.$$

Hence show that, if D contains an element f that is right cancellable for multiplication (in the sense that if $gf = hf$ then $g = h$), then D is in fact a ring.

5 Suppose that F is a field consisting of four elements a, b, c, d such that $a+b=b$ and $bc=c$. Show that there is only one possible Cayley table for multiplication, and find it. Deduce that $c(d+d)=b+b$. Hence deduce that there is only one possible Cayley table for addition, and find it.

Test paper 4

Time allowed: 3 hours
(Allocate 20 marks for each question.)

1 Prove that

$$G = \left\{ \begin{bmatrix} \cosh x & \sinh x \\ \sinh x & \cosh x \end{bmatrix} \mid x \in \mathbb{Q} \right\}$$

is a group under multiplication. Prove also that this group is isomorphic to the additive group $\mathbb{Q}$. Deduce that every finite subset of G is contained in a cyclic subgroup of G.

2 Let $E = \{1, 2, 3, 4, 5, 6\}$. By considering decompositions of permutations into disjoint cycles, or otherwise, determine all the permutations p on E such that $p^2 = (135)(246)$. Show that there is no permutation p on E such that $p^2 = (14)(25)(36)$.

3 Let $\alpha : R \to S$ and $\beta : R \to T$ be surjective ring morphisms. Prove that the following statements are equivalent:

(*a*) there is a ring morphism $\gamma : S \to T$ such $\gamma \circ \alpha = \beta$;

(*b*) $\operatorname{Ker} \alpha \subseteq \operatorname{Ker} \beta$.

Show that every ideal of the ring $\mathbb{Z}$ is of the form $n\mathbb{Z}$ for some $n \in \mathbb{Z}$. Deduce that if R is an infinite ring and $\alpha : \mathbb{Z} \to R$ is a surjective ring morphism then $R \cong \mathbb{Z}$.

4 Let A be a ring with identity element 1. Let $a \in A$ be such that

(*a*) a is not a zero divisor;

(*b*) $(\exists b \in A)\, ab = 1$.

By considering the element $aba - a$, or otherwise, show that a is invertible and that its inverse is b.

Suppose now that A in finite and that $x \in A$ is neither zero nor a zero divisor. If $A = \{x_1, \ldots, x_n\}$ show that the elements

$$xx_1, xx_2, \ldots, xx_n$$

are distinct, and deduce that $xy = 1$ for some $y \in A$. Conclude that x is invertible.

5 Let R be a ring. If there exists a least positive integer m such that $(\forall a \in R)ma = 0$ then R is said to be *of characteristic* m; and if no such integer exists then R is said to be *of characteristic zero.*

What are the characteristics of the rings $\mathbb{Z}$ and $\mathbb{Z}_n$?

If R has an identity element 1, prove that R is of characteristic m if and only if m is the least positive integer such that $m1 = 0$. Prove that an integral domain of characteristic non-zero is necessarily of characteristic p for some prime p. Deduce that a field is either of characteristic zero or of prime characteristic, and give an example of each.

If F is a field of prime characteristic p prove that

$$(\forall x, y \in F) \quad (x + y)^p = x^p + y^p.$$